W9-BAE-146

Mathematical Techniques for Biology and Medicine

WILLIAM SIMON
Professor of Biophysics
School of Medicine and Dentistry
The University of Rochester
Rochester, New York

Dover Publications, Inc., New York

Published in Canada by General Publishing Company, Ltd., 30 Lesmill Road, Don Mills, Toronto, Ontario.
Published in the United Kingdom by Constable and Company, Ltd., 10 Orange Street, London WC2H 7EG.

This Dover edition, first published in 1986, is an unabridged, corrected and enlarged republication of the edition published by The MIT Press, Cambridge, Mass., 1977. The latter was an expansion (for details, see "Preface to Paperback Edition") of the work first published by Academic Press, Inc., New York, 1972, under the title *Mathematical Techniques for Physiology and Medicine*. The author has added a new appendix (Appendix VII) to the Dover edition.

Manufactured in the United States of America
Dover Publications, Inc., 31 East 2nd Street, Mineola, N.Y. 11501

Library of Congress Cataloging-in-Publication Data

Simon, William, 1929–
Mathematical techniques for biology and medicine.

Reprint. Originally published: Cambridge, Mass. : MIT Press, c 1977.
Bibliography: p.
Includes index.
1. Biomathematics. 2. Medicine—Mathematics. I. Title. [DNLM: 1. Biometry. 2. Mathematics. 3. Medicine. 4. Physiology.
QH 323.5 S596m 1977a]
QH323.5.S52 1986 510′.24574 86-14208
ISBN 0-486-65247-5 (pbk.)

Contents

VII. Regulation and Oscillation

VIII. Diffusion

IX. The Theory of Blood Flow Measurement

X. Curve Fitting

NOTE: The "Preface to Paperback Edition" that follows was written in 1977; the "Preface," in 1972. The present Dover edition, 1986, contains a new appendix (Appendix VII) and a number of new corrections made directly in the text.

Preface to Paperback Edition

After this text was first published, I found it desirable to add Chapter XI describing the use of radioactive tracers. These techniques, which are of particular interest to cell physiologists and hematologists, may be studied immediately following Chapter V.

In some circumstances, the addition of this new chapter will make it difficult to cover the contents of the entire book in a one-semester course. One possible solution is to divide the class according to particular interests. Those students primarily interested in organ physiology should study Chapters IX and X, while those interested in cell physiology should study Chapter XI. Alternatively, as is the case at the University of Rochester, the material on numerical solutions found in Chapters VI and X is covered in another graduate-level course, and the material in Chapter XI is substituted in its place.

Two new appendices, containing additional problems and their solutions, have also been added to this edition.

I would like to acknowledge gratefully the assistance of Mrs. Christine Elsbeck in the preparation, editing, and assembly of Chapter XI and Appendices V and VI.

Preface

The prerequisite for this book is elementary calculus well-forgotten.

The book arose from the author's experience of four years as a "resident consultant" at a prominent medical school. During that time he became convinced that most medical students and life science students did not have sufficient mathematical background for research careers. Their usual approach to a compartmental problem or diffusion problem was to find someone else's treatment of a similar problem and hope that the procedures applied to their own. Alternatively, they consulted the local computing center where the lack of understanding was hidden behind the jargon of the computer world, but results were equally likely to be wrong.

Life science students often realize that they will need mathematical skills, and by taking at least elementary calculus and differential equations, attempt to acquire them. They are confronted, however, with a basic difficulty. The subjects treated in most college courses and the problem examples used are almost always from the fields of mechanics and electrical engineering. The biologist is forced to learn a new subject in what appears to him to be almost a foreign language. A mathematician's response to this is likely to be, "The principles are general. They should be able to apply them to biology." This response is both true and irrelevant. Perhaps if the student were to apply his knowledge to biological problems at the time of his undergraduate courses he would succeed in generalizing, but usually his first need for mathematics does not occur until graduate school. At that time he is faced with attempting to recall a subject learned in a foreign language never well understood, and required to translate that subject into a new context. The result often lies somewhere between frustration and disaster.

The strategy used in this book is one which has grown out of several years experience teaching mathematics to biologists. The book begins with the fundamentals of elementary calculus and proceeds quickly to the specific types of problems which are encountered in biological research. One may wonder why it is necessary to devote a full chapter, as well as the introduction to subsequent chapters, to material which should be familiar from elementary calculus. The answer is that almost invariably it is not sufficiently familiar. Typically, a graduate student in the life sciences remembers how to execute the process of taking a derivative, but he does not really know what a derivative is. He may remember it as the slope of a line but finds no way to connect this with the rate at which a tracer washes out of a compartment.

A number of unorthodox teaching techniques have been used in conjunction with this book. Students are encouraged to solve problems by guessing the forms of the solutions and trying them to see if they work. Hopefully, this will develop an intuitive understanding which is not acquired by formal solutions of problems. From the author's experience in teaching a course based on this technique, it has been interesting to note that students were reluctant at first to *guess* solutions to problems. Their previous training had always emphasized that in mathematics, *guessing* is not appropriate. However, when shown that the process is an educated guess rather than a blind one, they began to enjoy the game.

The choice of material included in this book, and in particular its limitations, is based on what is feasible to teach in one semester. Inherently, one must omit many topics that are the mathematician's bread and butter, such as complex numbers and the technique of separation of variables. These have been tried in the classroom situation and found to be too abstract. The Laplace Transform, on the other hand, seems to be acceptable and the tremendous power that it provides in the solution of problems is well worth the effort needed to understand its use.

Every effort has been made to reduce the number of new concepts, both biological and mathematical, with which the student must become familiar as he uses this book. The pedagogical principle is that it is better to teach a few new concepts well, by repetition where necessary, than to be encyclopedic. Proofs are rigorous wherever possible within the confines of a limited number of new ideas, and where rigor is not possible the results are made plausible by illustration, for example, the section on Fourier Series.

The biological examples used in this book are necessarily sketchy. Biological situations are so complicated that the best one can do in a volume of reasonable length treating a variety of subjects, is to establish principles and to supply references to sources which discuss specific problems in detail.

A course based on the material in this book has been taught for three years at The University of Rochester School of Medicine and Dentistry. Initially students complained that the course was too difficult. The results at the end of the semester proved contrary. The apparent difficulty stemmed largely from their being out of practice with mathematical manipulations. The book's first chapter is designed to correct this difficulty. In general, by the end of the semester students were able to comprehend literature which previously had been beyond their scope. This is the purpose of the course and of this book.

Mathematical Techniques for Biology and Medicine

I

Review of Differential Calculus

The mathematics of physiology serves two functions. The first is what most people usually think of as mathematics, a set of tools for providing numerical answers to problems. The second is that of a language by which concepts can be easily communicated and handled. As in any language, there is a certain basic vocabulary to be learned. This vocabulary consists largely of definitions that must be learned, just as in learning French, one learns that *poulet* means chicken. There is no logical way to derive this. It is a definition.

Some of the chapters begin with a mathematical introduction that describes the new mathematical concepts introduced in that chapter and which also contains their definitions. Try to think of these definitions as a new vocabulary and learn them as you would learn a vocabulary.

1. Dimensions

At many places throughout this book, the reader will observe that a dimensional equation has been written beneath the usual symbolic equation. It is hoped he will develop the habit of doing this himself at least twice in the solution of each problem: when the physical problem is stated in mathematical terms, and in the solution to the problem. Checking for dimensional balance should be a routine part of the solution of any problem. Remember, if it does not balance dimensionally, it is not correct. There are a few simple rules for manipulating dimensions.

Rule 1: Only quantities of like dimensions can be added or subtracted.

Rule 2: Dimensions multiply and divide in the same manner as numbers.

Example. To find the cost of 3 eggs plus 4 apples, given that E is the cost of eggs per dozen and A the cost of one apple,

$$C \text{ cents} = \frac{1}{4} \text{ dozen } E \frac{\text{cents}}{\text{dozen}} + 4 \text{ apples } A \frac{\text{cents}}{\text{apple}}$$

In the first term dozen cancels dozen, and in the second term apple cancels apple. Both are now in units of cents and can be added.

Rule 3: The easy way to convert from one set of dimensions to another is to write the conversion factor as a fraction whose value is 1. For example, to convert $12\frac{1}{2}$ feet to inches, construct the fraction 12 inches divided by 1 foot, and multiply this by $12\frac{1}{2}$ feet:

$$\frac{12 \text{ inches}}{1 \text{ foot}} \times 12\tfrac{1}{2} \text{ feet} = 150 \text{ inches}$$

The constructed fraction is equal to 1. Similarly, to convert 3 hours into seconds, construct the fraction 60 min divided by 1 hour, which is equal to 1, and a second fraction, 60 sec divided by 1 min, which is equal to 1. Multiply 3 hours by these two fractions, each of which is numerically one:

$$\frac{60 \text{ min}}{1 \text{ hour}} \quad \text{and} \quad \frac{60 \text{ sec}}{1 \text{ min}}$$

$$3 \text{ hours } \frac{60 \text{ min}}{1 \text{ hour}} \frac{60 \text{ sec}}{1 \text{ min}} = 10{,}800 \text{ sec}$$

Rule 4: Exponents must be dimensionless. When dimensioned quantities appear in exponents, they must combine with other dimensioned quantities so that the product or quotient is dimensionless. One cannot have a term such as 2^t where t is time. One can have

$$2^{t_1/t_2}$$

where the exponent is a ratio of two times (in the same units) and therefore dimensionless.

In dimensional equations we shall use the symbol * to indicate a dimensionless quantity.

When dimensional equations are used in this book they are usually written in terms of typical units rather than dimensional abbreviations. Other units having the same dimensionality can be substituted provided that the same substitution is made throughout the equation. For example,

$$C = \frac{Q}{V}$$

$$\frac{\text{moles}}{\text{liter}} = \text{moles}\,\frac{1}{\text{liter}}$$

$$\frac{\text{mmoles}}{\text{cm}^3} = \text{mmoles}\,\frac{1}{\text{cm}^3}$$

Except in this chapter, where other units are used for illustrative purposes, the units used in this book are metric.

2. The Concept of a Functional Relationship

The term "a function of" might be called a "depend-upon" relation. The postage required to send a package depends upon the distance the package must go and the weight of the package. A mathematician says that postage is a function of distance and weight. He indicates this relationship with the symbolism $P(D, W)$ and calls P a dependent variable because it depends upon the two independent variables D and W. The term "independent" implies that the independent variable can be chosen arbitrarily. You tell me a distance and weight and I will tell you, by some rule called the functional relationship, the postage required.

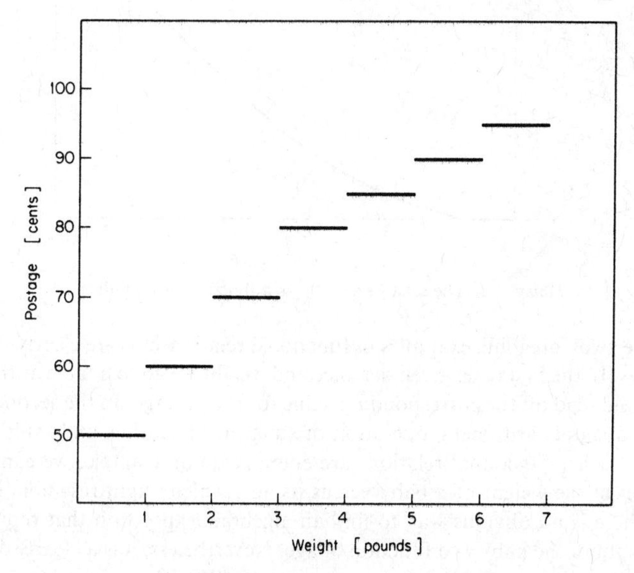

Figure 1-1. A graph of hypothetical postal rates for a package as a function of weight.

A dependent variable can be a function of one independent variable (Figure 1-1) or two independent variables (as in this case), or of many independent variables. Most of the problems we will encounter in the early parts of this book involve only a single independent variable. It is therefore easy to draw graphs that depict functional relationships. Conventionally, the independent variable is the horizontal axis of the graph.

Let us illustrate the concept of a functional dependence for some other simple cases. Consider the area A cm^2 of a circle of radius r cm. We know from plane geometry that

$$A = \pi r^2$$
$$\text{cm}^2 = * \text{ cm}^2$$

and we graph this functional relation $A(r)$ in Figure 1-2. If we wish to designate the area corresponding to a radius of 2 cm, we write A(2 cm), by which it is understood that the independent variable will take on the value 2 cm even though the independent variable r is not specifically written in this notation. We might even leave out the symbol cm if this is made clear by context, and simply write $A(2)$.

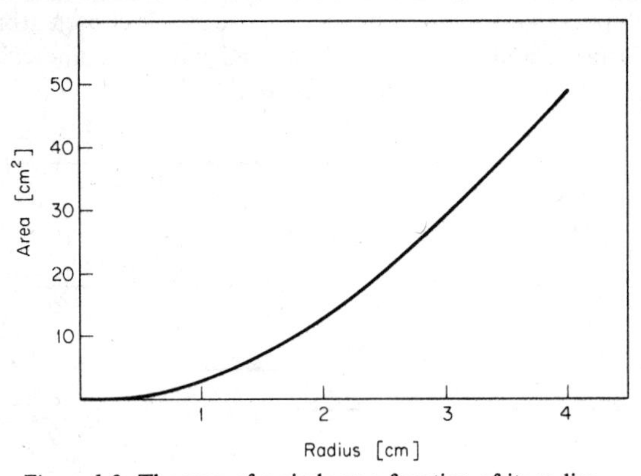

Figure 1-2. The area of a circle as a function of its radius.

The two foregoing examples of functional relationships are clearly defined by rules. In the first case, given distance and weight we go to a table at the post office and read off the corresponding value for the postage. In the second case, we do a simple arithmetic operation of squaring the radius and multiplying it by π. Other functional relations are empirical. For example, we can make a graph of the weight of a baby versus its age, which might resemble Figure 1-3. There is no obvious way to find an algebraic expression that represents the weight of the baby as a function of age. Nevertheless, it is a clearly defined number and it is perfectly appropriate to write W(age).

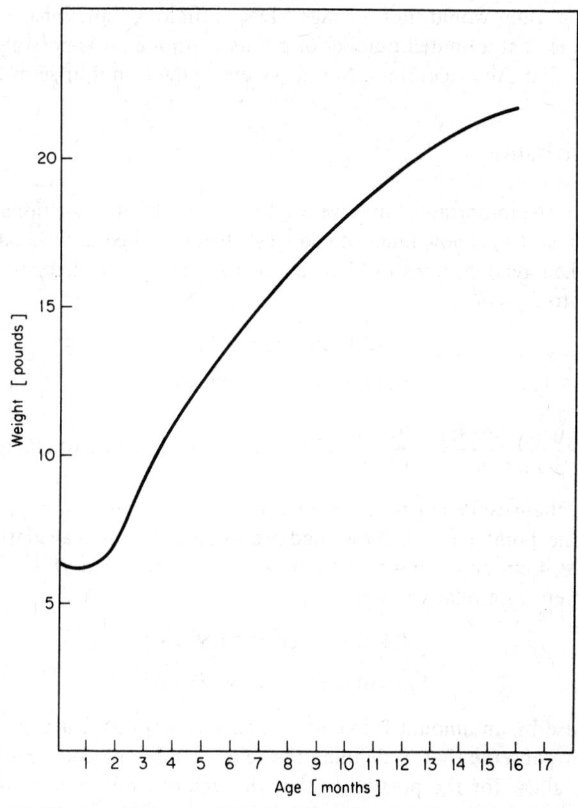

Figure 1-3. The hypothetical weight of a baby plotted against his age in months.

The last two examples, the area of the circle and the weight of the baby, have a property that mathematicians refer to as being *continuous functions.* The term "continuous" means simply a smooth relationship. Given two values of the independent variable, the radius, for example, that are very close to each other, the two corresponding values of the dependent variable will then also be very close to each other. This is not true, for example, of the postage required to send a package a specified distance, since by post office rules a package that weighs a trifle more than 6 pounds goes at the 7-pound rate, whereas a package that weighs a trifle less than 6 pounds goes at the 6-pound rate. No matter how close to 6 pounds each of these packages becomes, the rates do not get closer together but remain discretely at the 6-pound rate and the 7-pound rate. Thus, postage is a discontinuous function even though it is continuous between say, 6.01 pounds and 6.99 pounds, where the rate would not change. Discontinuous functions are usually discontinuous at a limited number of points. Although it is possible to define functions that are discontinuous everywhere, they do not arise in this book.

3. The Derivative

One of the important things we want to know about a functional relationship such as $A(r)$ is how much A changes when r changes a little bit (in other words, their relative rates of change). If, for example, r changes by 0.1 cm from 3.0 to 3.1 cm,

$$A(3.1)\ \text{cm}^2 = 30.19\ \text{cm}^2$$

$$A(3.0)\ \text{cm}^2 = 28.27\ \text{cm}^2$$

$$\frac{\text{change in } A}{\text{change in } r} = \frac{1.92\ \text{cm}^2}{0.1\ \text{cm}} = 19.2\ \text{cm} = \text{relative rate of change}$$

Thus, A changes 19 cm times as fast as r for a small change in r located around the point $r = 3.0$. Now, had we done the same calculation with r starting at 4 cm and going to 4.1 cm, we would find that A would change at a different rate relative to r:

$$A(4.1) = (4.1)^2 = 52.80\ \text{cm}^2$$

$$A(4.0) = (4.0)^2 = 50.25\ \text{cm}^2$$

in this case by an amount 2.55 cm^2 or 25.5 cm times as fast as r. So that in general we observe that in defining the relative rates of change of A and r we must allow for the possibility that this relative rate is not uniform but changes as r changes. We can easily draw a picture by constructing a little triangle around the point $r = 3$ cm on the graph of r versus A in which the horizontal leg of the triangle is the change in r and the vertical leg of the triangle the change in A (Figure 1-4). The relative rates of change are then given by the ratios of the two sides of the triangle, and we see that if we construct the triangle in a variety of places along the graph, the ratio of the two sides will change. Our original question was, how fast does A change relative to r around the point $r = 3$? The quantity we have actually calculated is not quite this. It is, in fact, a sort of average of this rate between the point $r = 3$ cm and the point $r = 3.1$ cm. We would perhaps have gotten a slightly different answer if we had calculated the relative rates of change between $r = 3$ cm and $r = 3.01$ cm, which we proceed to do:

$$\frac{A(3.01) - A(3.00)}{3.01 - 3.00} = \frac{28.46 - 28.27}{0.01} = 19.0\ \text{cm}$$

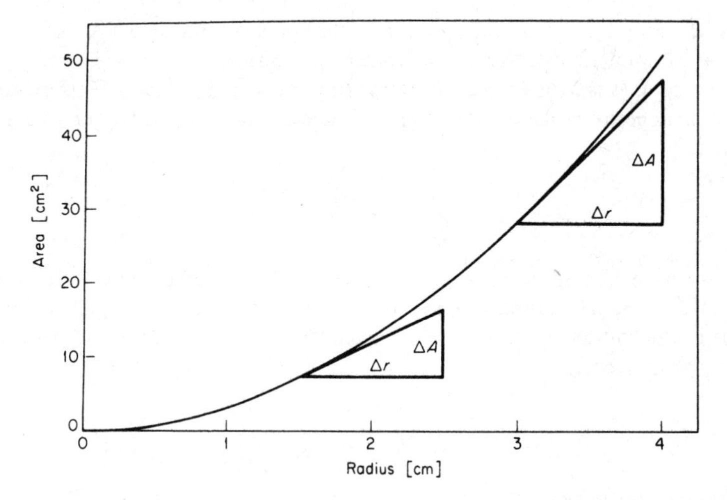

Figure 1-4. The definition of a derivative in terms of tangents to a line.

We note that this relative rate is not much different from the one we calculated previously, but it is slightly different and that in order to define the precise rate of change around the point $r = 3$ we should really do the calculation with an extremely small change in r.

The process we describe here should be studied carefully, as it is fundamental to the rest of this book.

In order to find the rate of change of A relative to the rate of change of r, we calculate the value of A at the point r and at an adjacent point $r + \Delta r$ where Δr is understood to be very small and will eventually be made infinitesimal or, to use the conventional terminology, Δr will be allowed to go to zero. We have already done the calculation of $A(r)$, but let us repeat it in a symbolic fashion rather than with numbers. The calculation of $A(r + \Delta r)$ proceeds along exactly the same line except that $r + \Delta r$ takes the place of r:

$$A(r) = \pi r^2$$

$$A(r + \Delta r) = \pi(r + \Delta r)^2 = \pi(\Delta r)^2 + 2\pi r\,\Delta r + \pi r^2$$

We now calculate the difference ΔA between the two areas thus obtained and divide this difference by the quantity Δr, just as we did in the numerical example above,

$$\frac{\Delta A}{\Delta r} = \frac{A(r + \Delta r) - A(r)}{\Delta r} = \pi 2r + \Delta r\pi$$

The quantity thus obtained, the ratio of the two small quantities ΔA and Δr, is called a differential. We now consider what happens as Δr gets very, very small, or as it is formally stated, in the limit as Δr goes to zero. This procedure is done so frequently that a special notation, called a *derivative*, is used for it:

$$\frac{dA}{dr} = \operatorname*{limit}_{\Delta r \to 0} \frac{A(r + \Delta r) - A(r)}{\Delta r} = 2\pi r$$

We have worked out in somewhat tedious detail the derivative of a very simple functional relationship. Let us now define it formally, recognizing that this is a definition. *The derivative of a function $f(x)$ relative to its independent variable is given by*

$$\frac{df(x)}{dx} = \operatorname*{limit}_{\Delta x \to 0} \frac{f(x + \Delta x) - f(x)}{\Delta x} \tag{1.1}$$

when Δx goes to zero.

In most beginning calculus texts the derivative is first defined as the slope of the hypotenuse of one of the little triangles in Figure 1-4. Although this definition has a certain intuitive appeal, for the purposes of this book the formal definition is far more satisfactory.

When one wishes to indicate the value of a derivative at a particular value of the independent variable, this value is enclosed in parentheses. Thus the value of df/dx evaluated at x equal to 4 is $df(4)/dx$.

Returning to the area example, we have

$$\frac{dA(3)}{dr} = 2\pi 3 \text{ cm} = 18.8 \text{ cm}$$

Note that the dimensions of the derivative dA/dr are the same as the dimensions of A/r. This is a general rule. Dimensionally, df/dx is the same as f/x.

The derivative of a function as defined above is the rate of change of the function f relative to the rate of change of the independent variable x for infinitesimal changes in x. Frequently we want to know how much f changes for a small but not infinitesimal change in x. To a very good approximation, indicated by the symbol $\approx$, this change is given by

$$\Delta f \approx \Delta x \frac{df}{dx} \tag{1.2}$$

The error in this approximation is the difference between the true value of $f(x + \Delta x)$ and the triangular approximation we find by drawing a tangent line to the curve of f versus x (Figure 1-5).

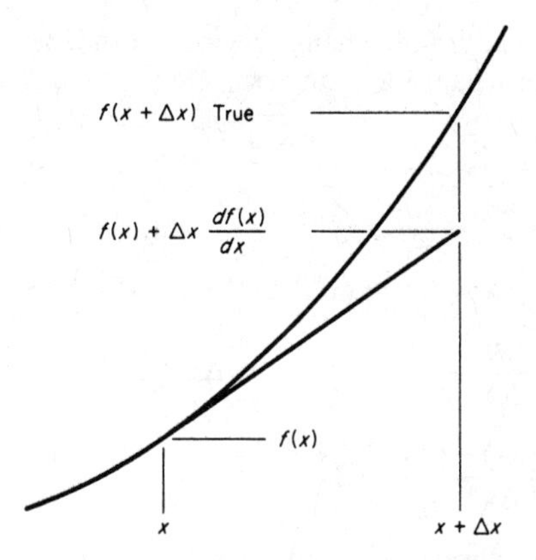

Figure 1-5. How true rate of change of a function is approximated by Δx multiplied by its derivative.

Using this approximation, let us find the approximate value of $A(3.1)$ given that $A(3.00)$ is 28.27 cm^2:

$$\frac{dA}{dr} = 2\pi r$$

$$\frac{dA(3)}{dr} = 2\pi 3.0 = 18.84 \text{ cm}$$

$$\Delta r = 0.1 \text{ cm}$$

so that

$$A(3.1) \approx A(3.0) + \Delta r \frac{dA(3)}{dr} = 28.27 + 0.1 \times 18.8$$

$$= 30.15 \text{ cm}^2$$

The exact value is 30.19 cm^2. Had we chosen a smaller interval Δr, the approximation would have been better.

In the previous example we have calculated the derivative of

$$A = \pi r^2$$

To calculate the derivative of any function we proceed along identical lines, calculating symbolically the value of the function at some value x of the independent variable and at some slightly different value $x + \Delta x$ of the independent variable. We then compute the difference between the values thus calculated, divide by the change Δx of the independent variable, and take the limit as this change goes to zero. Thus to find the derivative of

$$y = x^3$$

proceed as follows.

$$y(x) = x^3$$

$$y(x + \Delta x) = (x + \Delta x)^3 = x^3 + 3x^2\,\Delta x + 3x(\Delta x)^2 + (\Delta x)^3$$

$$\frac{y(x + \Delta x) - y(x)}{\Delta x} = 3x^2 + 3x\,\Delta x + (\Delta x)^2$$

$$\operatorname*{limit}_{\Delta x \to 0} \frac{y(x + \Delta x) - y(x)}{\Delta x} = 3x^2$$

In principle, we can always calculate derivatives in this way. But this is not the way mathematicians like to do things. Instead they derive a set of rules which, though tedious to develop, then make future problems simpler. We therefore state a set of rules that are useful in calculating derivatives with the hope that the reader will either remember (from his elementary calculus) how these are derived, or will look at Appendix I.

Rule: The derivative of a constant is zero. This rule requires a moment's explanation. A constant can be a function of a variable. It just happens to be a function that never changes its value. Thus, for example, one might say that the number of hours in a calendar day is a function of the day of the year; but it is a function that never changes from the value 24. Therefore the difference between its value calculated at one time (the independent variable) and at another time, slightly different, is zero. Therefore its derivative is zero.

Rule: The derivative of a constant multiplied by a function is the constant multiplied by the derivative of the function.

Rule: The derivative of the sum of two functions is the sum of their individual derivatives:

$$y = f(x) + g(x)$$

$$\frac{dy}{dx} = \frac{df}{dx} + \frac{dg}{dx} \tag{1.3}$$

Rule: The derivative of a product of two functions is the first multiplied by the derivative of the second plus the second multiplied by the derivative of the first:

$$y = f(x)g(x)$$

$$\frac{dy}{dx} = f(x)\frac{dg(x)}{dx} + g(x)\frac{df(x)}{dx} \tag{1.4}$$

Rule: The derivative of the quotient of two functions

$$y = \frac{f(x)}{g(x)}$$

is the denominator multiplied by the derivative of the numerator minus the numerator multiplied by the derivative of the denominator, all divided by the square of the denominator:

$$\frac{dy}{dx} = \left(g(x) \frac{df(x)}{dx} - f(x) \frac{dg(x)}{dx} \right) \frac{1}{g^2(x)} \tag{1.5}$$

Note as a special case of this rule that the derivative of a reciprocal

$$y = \frac{1}{f(x)}$$

is

$$\frac{dy}{dx} = -\frac{1}{f^2(x)} \frac{df}{dx}$$

Rule: The derivative of $y = x^n$ with respect to x is

$$\frac{dy}{dx} = nx^{n-1} \tag{1.6}$$

These rules are used separately or in combination to reduce a function whose derivative is required to a combination of functions whose derivatives are known. Thus, for example, to find the derivative of the function

$$y = \frac{x^2 - 2x}{x - 1}$$

we proceed in the following way, using the quotient rule

$$\frac{dy}{dx} = \frac{1}{(x-1)^2} \left[(x-1) \frac{d}{dx} (x^2 - 2x) - (x^2 - 2x) \frac{d}{dx} (x - 1) \right]$$

$$= \frac{1}{(x-1)^2} [(x-1)(2x-2) - (x^2 - 2x)]$$

$$= 2 - \frac{x(x-2)}{(x-1)^2}$$

Sometimes we have a function that depends upon a second function: A is a function of B and B is a function of C. We ask for the derivative of A

with respect to C; in other words, how much does A change if C changes a little bit? To find this, we must go through an intermediate step. We know that for a small change in C, B will change by approximately

$$\Delta B = \Delta C \frac{dB}{dC}$$

We know that for a change in B, A will change as follows:

$$\Delta A = \Delta B \frac{dA}{dB}$$

Thus if we substitute the change in B in the equation above, we find the result that the rate of change of A relative to C is the product of the rates of change of A relative to B and B relative to C:

$$\Delta A = \Delta C \frac{dA}{dB} \frac{dB}{dC}$$

$$\frac{dA}{dC} = \frac{dA}{dB} \frac{dB}{dC} \tag{1.7}$$

This is called the *chain rule*. It is easily remembered by considering the derivatives as fractions in which the two dB quantities cancel.

4. The Function That Is Its Own Derivative

In general, the derivative of a function is a different function. Thus the derivative of x^3 is $3x^2$. The derivative of

$$y = \sqrt{x}$$

is

$$\frac{dy}{dx} = \frac{1}{2\sqrt{x}}$$

and so on. There is, however, one rather peculiar function and its usefulness is the result of its peculiarity. It is its own derivative. That function is

$$y = e^x, \qquad e \approx 2.7182 \tag{1.8}$$

$$\frac{dy}{dx} = e^x \tag{1.9}$$

Why e^x has this property, and why this strange number occurs, is explained in Appendix I. It is based on the representation of e^x by a series with an infinite number of terms, which is

$$e^x = 1 + \frac{x}{1} + \frac{x^2}{1 \cdot 2} + \frac{x^3}{1 \cdot 2 \cdot 3} + \frac{x^4}{1 \cdot 2 \cdot 3 \cdot 4} + \cdots$$

When this series is differentiated by differentiating each term

$$\frac{d}{dx} e^x = 0 + \frac{1}{1} + \frac{2x}{1 \cdot 2} + \frac{3x^2}{1 \cdot 2 \cdot 3} + \frac{4x^3}{1 \cdot 2 \cdot 3 \cdot 4} + \cdots$$

the resulting series is seen to be identical to the series for e^x.

Accepting the fact that e^x is its own derivative, we proceed, by means of the chain rule, to find the derivative of e to a function of x. Let p be a function of x

$$y = e^{p(x)}$$

$$\frac{dy}{dp} = e^{p(x)}$$

$$\frac{dy}{dx} = \frac{dy}{dp}\frac{dp}{dx} = e^{p(x)}\frac{dp}{dx} = y\frac{dp}{dx}$$

and in the special case where the function is a constant, say alpha, multiplied by x, we have

$$y = e^{\alpha x}$$

$$\frac{dy}{dx} = e^{\alpha x}\frac{d(\alpha x)}{dx} = \alpha e^{\alpha x} = \alpha y$$

The function $y = e^{\alpha x}$ occurs so frequently that we would like to describe its properties in considerable detail.

First, let us make a plot of

$$y = e^{\alpha x} \tag{1.10}$$

on a conventional rectangular graph (Figure 1-6).

To illustrate the point that $\alpha e^{\alpha x}$ is the derivative of $e^{\alpha x}$, let us take two points from the graph and numerically compute the derivative.

Let us choose $\alpha = 1.5$, $x = 1.4$, $\Delta x = 0.1$; at

$$x = 1.4, \qquad \alpha x = 2.10, \qquad e^{\alpha x} = 8.0$$

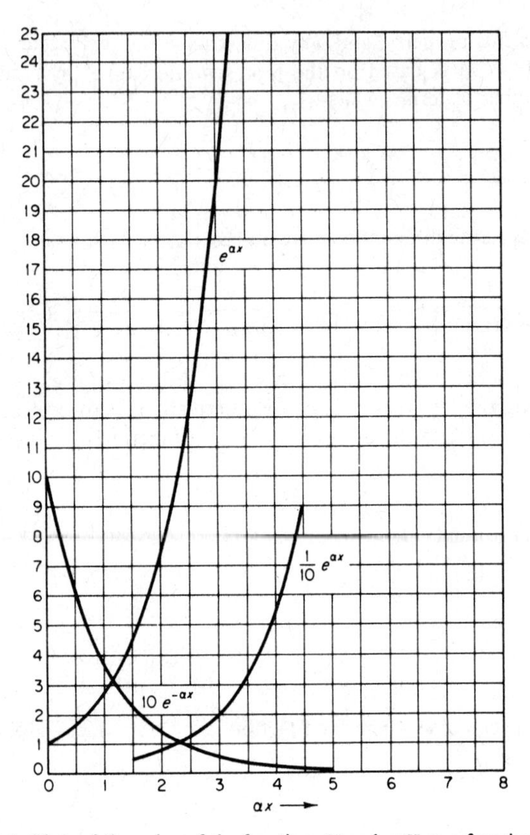

Figure 1-6. Plots of the value of the function $e^{\alpha x}$ and $e^{-\alpha x}$ as a function of αx.

and at

$$x + \Delta x = 1.5, \qquad \alpha x = 2.25, \qquad e^{\alpha x} = 9.3$$

$$\frac{\Delta y}{\Delta x} = \frac{y(x + \Delta x) - y(x)}{\Delta x} = \frac{9.3 - 8.0}{0.1} = 13$$

Using the relation

$$\frac{dy}{dx} = \alpha e^{\alpha x} = 1.5e^{2.10} = 12.0$$

We see that finding the derivative numerically does not quite yield perfect agreement between dy/dx and $\Delta y/\Delta x$. However, this is a result of taking a finite step in x and of reading the graph inaccurately. Had the step been taken much smaller (that is, had it been possible to do so from this graph), the agreement would have been much better. In fact, let us do so, by expanding the graph around the point $\alpha x = 2.1$ in Figure 1-7 and using a Δx of 0.02:

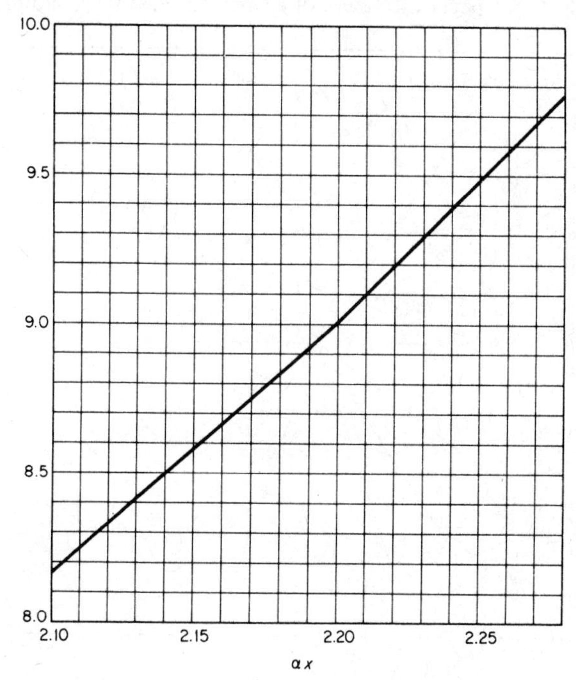

Figure 1-7. An expanded scale plot of $e^{\alpha x}$ as a function of αx.

at

$$x = 1.40, \qquad \alpha x = 2.10, \qquad e^{\alpha x} = 8.16$$

and at

$$x = 1.42, \qquad \alpha x = 2.13, \qquad e^{\alpha x} = 8.41$$

$$\frac{\Delta y}{\Delta x} = \frac{y(x + \Delta x) - y(x)}{\Delta x} = \frac{0.25}{0.02} = 12.5$$

$$\frac{dy}{dx} = \alpha e^{\alpha x} = 1.5e^{2.10} = 12.24$$

We note that we get considerably better agreement between the value of the derivative and the numerical differential.

The function $y = e^{\alpha x}$ occurs so often that special graph paper has been devised which plots it as straight lines by distorting the vertical scale. This paper is called *semilogarithmic* graph paper. In Figure 1-8 we have plotted on semilogarithmic paper the values of $e^{\alpha x}$ for various values of alpha.

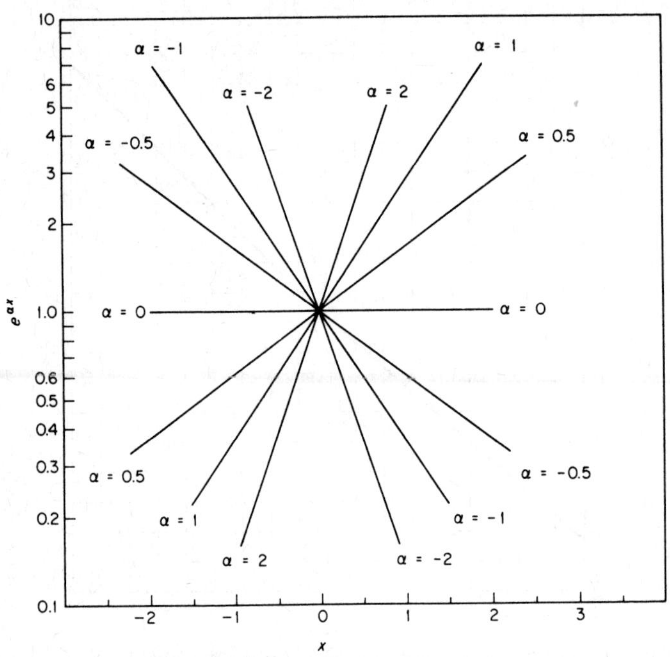

Figure 1-8. $e^{\alpha x}$ plotted semilogarithmically as a function of x for various values of α.

5. Differential Equations

The statement that e^x is its own derivative can be written as an equation

$$\frac{dy}{dx} = y \tag{1.11}$$

This kind of equation is one of a great class of equations that are encountered in physical problems called *differential equations.* They should, perhaps, be called *derivative* equations. However, tradition rules here, and we shall retain the terminology differential equation. The differential equation (1.11), since it is the definition of e^x, has the solution

$$y = e^x \tag{1.12}$$

It happens, however, that it also has other solutions. In fact, any constant multiplied by e^x is also a solution. Let A be an arbitrary constant.
Let

$$y = Ae^x \tag{1.13}$$

Then

$$\frac{dy}{dx} = A\frac{de^x}{dx} = Ae^x = y \tag{1.14}$$

We make use of this property to fit solutions of the type Ae^x to specified conditions of the physical problem. If, for example, we have the relations

$$\frac{dy}{dx} = y \quad \text{and} \quad y(0) = 3$$

we can choose $A = 3$ to satisfy the "initial condition" of $y(0) = 3$.

$$y = 3e^x = \frac{dy}{dx} \quad \text{and} \quad y(0) = 3$$

Suppose the differential equation is

$$\frac{dy}{dx} = ky$$

Our previous discussion suggests a solution

$$y = Ae^{kx}$$

Let us try this solution by differentiating it and inserting the derivative into the differential equation.

$$\frac{dy}{dx} = Ake^{kx} = ky$$

We find that, indeed, it does satisfy the differential equation. If we add the condition $y(2) = 4$,

$$4 = Ae^{k2}$$

we find

$$A = \frac{4}{e^{2k}}$$

which, given k, we can evaluate.

Plotted on semilogarithm paper, as in Figure 1-9, the relation $y = Ae^{kx}$ also yields a straight line. The value of the vertical coordinate at $x = 0$ is A, in this case 2.6. The slope of the line is related to k. The easy way to measure this slope is to find the range of x required to double or halve the value of y. For the line shown in Figure 1-9, y decreases by half in the interval between

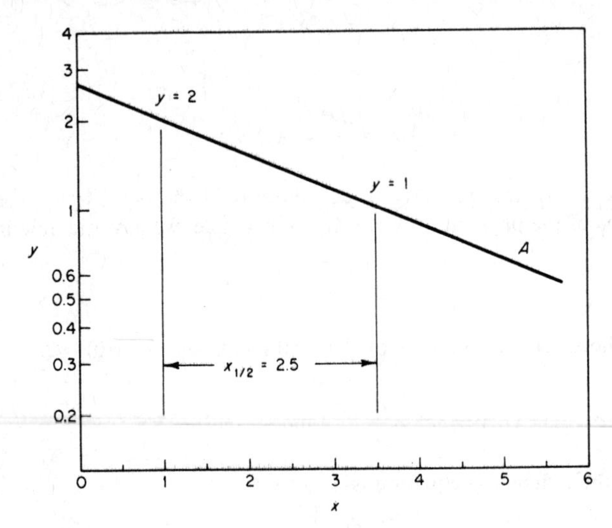

Figure 1-9. The method of estimating the half period of decay of an exponential from a semilogarithmic plot.

$x = 1$ and $x = 3.5$. Thus, the half range $x_{1/2}$ is 2.5, and since

$$e^{-0.69} = \tfrac{1}{2}$$

$$kx_{1/2} = -0.69 \tag{1.15}$$

$$k = \frac{-0.69}{2.5} = -0.276$$

Note that we could have chosen any two convenient points on the graph to determine $x_{1/2}$. We chose the points at which y was 2 and 1. We could equally well have chosen y equal to 1.5 and 0.75. They would have yielded the same value for k. Had the line sloped upward, we would have used the doubling range x_2 and the relation

$$e^{+0.69} = 2, \qquad kx_2 = 0.69 \tag{1.16}$$

The foregoing introduction has been intentionally brief. If the reader finds it inadequate, he should refer to his college calculus text or to either (18) or (23).

Problems

1. If $W = G(x)$, define the derivative

$$\frac{dW}{dx} = \text{limit} \quad ?$$

2. $W = x^2 - 2x + C$ where C is a constant; find

$$W(2) \qquad \text{and} \qquad \frac{dW(2)}{dx}$$

3. Given

$$U(x) = W(x)V(x) \qquad S(x) = \frac{V(x)}{W(x)}$$

$$W(2) = 3 \qquad \frac{dW(2)}{dx} = 4$$

$$V(2) = 4 \qquad \frac{dV(2)}{dx} = 2$$

find

$$\frac{dU(2)}{dx} \qquad \text{and} \qquad \frac{dS(2)}{dx}$$

4. If $y(x)$ is a smooth function of x and $y(3) = 4$ and $y(3.1) = 4.3$, find the approximate value of $dy(3)/dx$.

5. Given

$$\frac{dy(3)}{dx} = 2 \qquad y(3) = 12$$

find the approximate value of $y(3.1)$.

6. If $u = f(x)$ and $V = g(u)$, find dV/dx.

7. In Problem 6 let

$$f(x) = x^4 + 2, \qquad g(u) = u^2$$

find dV/dx by the chain rule and also by substituting

$$V = g(u) = u^2 = x^8 + 4x^4 + 4$$

8. Given

$$\frac{dC(t)}{dt} = kC(t), \qquad C(0) = 2 \text{ moles/liter}$$

find $C(t)$.

9. Given

$$\frac{dC(t)}{dt} = kC(t), \qquad C(1) = 2 \text{ moles/liter}$$

find $C(t)$.

10. Given

$$y = Ae^{-kt}, \qquad y(4 \text{ sec}) = \tfrac{1}{2}y(2 \text{ sec})$$

find k.

11. Find the decay constant α of $y = Ae^{-\alpha t}$ for the following data:

t [sec]	y
1	0.0096
2	0.0048
3	0.0024
4	0.0012

12. Derive the relations corresponding to Eqs. (1.15) and (1.16) for the 1/10 interval $\tau_{1/10}$ or 10 times interval τ_{10}.

13. If the following results occurred in problems, what must the dimension of K have been in each case?

(a) e^{-kt^2}, $\quad t$ in seconds

(b) $C = \dfrac{F}{K}t$, $\quad F \dfrac{\text{liters}}{\text{sec}}$, $\quad t$ sec, $\quad C \dfrac{\text{moles}}{\text{liter}}$

(c) $R = (F + K)e^{-(F/V)t}$, $\quad F \dfrac{\text{liter}}{\text{sec}}$

14. The velocity of light is 3 times 10^{10} or 30,000,000,000 cm/sec. Find the velocity of light in furlongs per fortnight given that a furlong is one-eighth of a mile, a mile is 5280 feet, a foot is 12 inches, an inch is 2.54 cm, and a fortnight is 2 weeks. Use the unit conversion method in Section 1.

15.

$$\frac{dC}{dt} = \frac{R}{V} - \frac{F}{V}C$$

Write a dimensional equation corresponding to the foregoing if C is a concentration, t time, R a rate of generation of a substance, V a volume, F a rate of liquid flow.

16.

$$\frac{d^2C}{dt^2} + B\frac{dC}{dt} + DC = E$$

Write a dimensional equation for the foregoing, given that C is a concentration and t is time.

17. Find the derivatives of the following.

(a) $y = x^4 + 3x^3 + x + 2$ (b) $y = \dfrac{x^3 - x}{x^3 + x}$

(c) $y = e^{x^2}$ (d) $y = xe^{\alpha x}$

If the reader has difficulty with these problems, additional review of elementary calculus is indicated.

II

How Differential Equations Arise

Suppose one had never seen a spider web, but came upon the pattern in Figure 2-1 woven out of silk. If the spider assured us that he had followed the same design throughout his entire web out to a radius of 20 cm, we could easily infer what a spider web looks like. In the same way, many physical chemical, and biological processes can be described by a rule that tells us what happens in part of the process. If we are assured that the same rule always applies throughout the process, it is possible to derive the behavior of the entire process. Let us take a simple example.

Example. Suppose we are assured that following an injection of an indicator or tracer into an organ at time t zero, 1% of the tracer is lost through the outflow of the organ in each second of time. Let the initial amount of tracer be 0.300 moles; then at the end of 1 sec we have 0.297 moles, at the end of 2 sec we have about 0.294 moles, and so on. Thus we are able to calculate the amount of tracer remaining in the compartment for any time, however long, and can tabulate it as follows.

Time [sec]	Q [moles]
0	0.300
1	0.297
2	0.294
3	0.291
4	0.288

It is apparent that we need to specify three things: the rule that governs a process over a short interval of time, in this case, 1 sec; the assurance that the same rule applies over a long interval of time, and the initial amount of tracer in the organ.

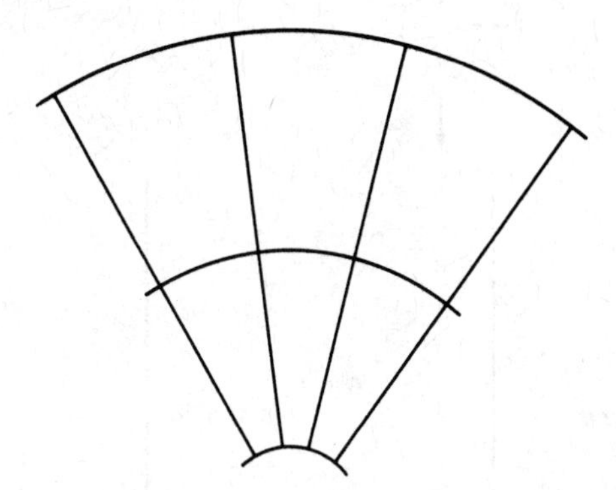

Figure 2-1. Part of a hypothetical spider web.

Suppose, however, we change the rule; make it 2% loss of tracer in each second. Using the method just described, it would be necessary to repeat the entire calculation. The object of the differential equation formulation of a problem is to solve it in so general a way that specific results, for example, the amount of tracer left at 20 sec, can be obtained from relatively simple computation, taking advantage of the fact that there exist tables of certain common mathematical functions, such as the value of e^x, that would enter into the solution of this problem.

The remainder of this chapter discusses a number of ways in which differential equations arise in biological problems. These equations will not be formally solved at this stage; rather, we will simply state the solutions and verify them by differentiating and trying them in the differential equations. Attention will be devoted principally to the steps that occur between the physical description of a problem and its mathematical description.

The previous example is one of compartmental dilution, which arises in a number of differing circumstances in biology but in its simplest form describes the washout of tracers or indicators, such as dyes or radioactive elements, from vascular or cellular systems.

1. The One-Compartment Dilution Process

The idealized one-compartment dilution problem consists of a single continuously mixed chamber through which a fluid is flowing at a constant rate. One can visualize this, for example, as a bucket, with a hole in the bottom from which fluid is flowing, and a garden hose keeping the bucket filled to a fixed level (Figure 2-2). Consider a bucket of volume 10 liters

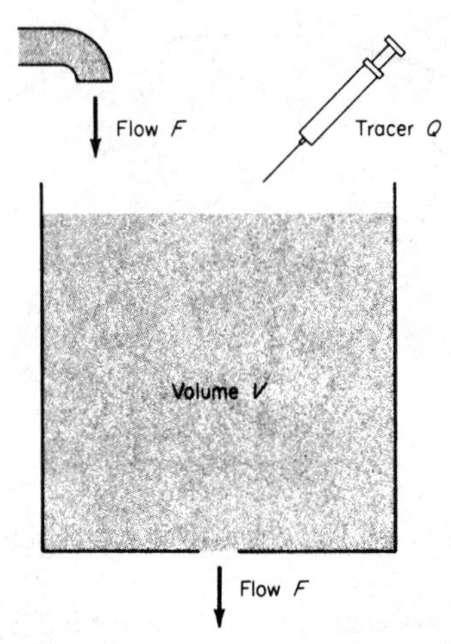

Figure 2-2. The basic one-compartment continuous dilution problem.

through which water is flowing at a rate of 2 liters/min. At some arbitrary time, which we call $t(0)$, 3 moles of a tracer, a colored dye, for example, is injected into the bucket. We assume for simplicity that the dye is injected instantly and that it is instantly mixed, so that at time $t(0)$ the concentration will be 0.3 moles/liter. This concentration will immediately begin to fall, since dye is being lost in the effluent while water is replacing the lost volume of solution. Thus, at very long times, which we designate $t(\infty)$, all of the dye has been flushed out of the bucket. To find the concentration of dye at any time following its injection, we must introduce the concept of instantaneous description of a continuous process. The rate at which dye is washed out of the bucket changes with time. It is initially at its maximum value because the concentration is at a maximum immediately following injection. As dye is lost, the concentration decreases, and since the outflow is at a fixed rate, the rate at which dye is lost decreases with time. We must therefore describe the process in a way that takes into account the lessening concentration with increasing time. Let us pick some arbitrary time, which we will call t, and describe what happens to the concentration of dye between time t and a slightly greater time t plus Δt. Here Δt represents some small but nonzero interval of time. Thus one might choose t as 150 sec and Δt as 1 sec, so that we are trying to describe the change in concentration that occurs between 150 and 151 sec. During this time concentration changes only by a very small fraction of its value. It goes from a value $C(t)$ to $C(t + \Delta t)$. These notations mean the concentration at time t and at time $t + \Delta t$.

The loss ΔQ of tracer that occurs during the interval Δt depends upon its concentration C during Δt. By choosing Δt small we can guarantee that, during Δt, C changes by only a small fraction of its value and we can calculate ΔQ from the value of C at the beginning of Δt.

During the time interval Δt the amount of solution that flows out of the tank is $F\,\Delta t$ liters, flow rate F multiplied by the time interval. This fluid will carry with it a quantity of dye ΔQ equal to the concentration at time t (according to the approximation above) multiplied by the amount of solution that has been lost during that time. Thus

$$\Delta Q = -\Delta t \qquad \mathrm{F} \qquad \mathrm{C}$$

$$\text{moles} = \text{min} \qquad \frac{\text{liter}}{\text{min}} \qquad \frac{\text{moles}}{\text{liter}}$$

The minus sign arises because the amount of tracer in the bucket is being reduced by the washout. Note that ΔQ is a quantity, moles, not a concentration, which would be moles per liter. To find how the *concentration* will change in the same time, we note that the concentration is equal to the quantity of dye in the bucket divided by the volume of the bucket. Thus the change in concentration equals the change in quantity of dye divided by the volume:

$$\Delta C = -\Delta t \qquad F \qquad C \qquad \frac{1}{V}$$

$$\frac{\text{moles}}{\text{liter}} = \text{min} \qquad \frac{\text{liters}}{\text{min}} \qquad \frac{\text{moles}}{\text{liter}} \qquad \frac{1}{\text{liters}}$$

Note that ΔC is not zero. It is a small percentage of C but it is not zero. It is also slightly in error because we calculated it from the value of C at the beginning of Δt, but we shall now make it as exact as we wish by making Δt very much smaller than 1 sec. Note that as we do so, ΔC will become smaller, since from the foregoing equation it is proportional to Δt. However, the ratio between ΔC and Δt will not change except for a very small amount due to using $C(t)$ for the purpose of calculating ΔC. Divide both sides of the equation by Δt and observe that the left side is a quotient of two very small quantities in the form of the differential described in Chapter I. We then take the limit as Δt goes to 0 and replace the quotient of the two small quantities by the notation

$$\frac{dC(t)}{dt} = \lim_{\Delta t \to 0} \frac{\Delta C}{\Delta t} = -\frac{FC(t)}{V}$$

$$\frac{\dfrac{\text{moles}}{\text{liter}}}{\text{min}} = \frac{\dfrac{\text{liter}}{\text{min}}\,\dfrac{\text{moles}}{\text{liter}}}{\text{liters}} \tag{2.1}$$

This is our first differential equation. It is one in which the derivative of C is proportional to C and is therefore of the type discussed in Chapter I, Section 4.

Its solution is

$$C(t) = C(0)e^{-(F/V)t}$$

$$\frac{\text{moles}}{\text{liter}} = \frac{\text{moles}}{\text{liter}} \tag{2.2}$$

where the exponent has units of

$$\frac{F}{V}t = \frac{\frac{\text{liters}}{\text{min}}}{\text{liters}}\,\text{min}$$

and is therefore dimensionless, as it must be.

$C(0)$ at this stage is an arbitrary constant, a number to be determined later. This solution has not been derived but can be verified as follows:

Differentiate the solution with respect to time and insert the derivative into the differential equation. Since the derivative of $e^{-(F/V)t}$ is $-(F/V)$ $e^{-(F/V)t}$,

$$\frac{dC(t)}{dt} = -\frac{F}{V}C(0)e^{-(F/V)t} = -\frac{F}{V}C(t) \tag{2.3}$$

We note that the differential equation is now satisfied. However, we must still satisfy the initial condition that the concentration at time 0 must be 0.3 mole/liter. Since any $C(0)$ satisfies the differential equations, it can be chosen to be 0.3 mole/liter. This is now the complete solution to our problem.

Before discussing the properties of this solution, let us review the strategy used to obtain it.

1. We specified the physical conditions of the problem, in this case a tank of water of volume V through which there is a constant flow at a rate F liters/min. The volume of the tank was assumed constant. The tracer was injected in a very short interval of time at time 0. The tank was assumed to be kept continuously stirred so that it had the same concentration at all points.

2. We calculated how much the concentration would change in some small interval of time called Δt, which lies between the time t and the time $t + \Delta t$, and although we were calculating a *change* in concentration that occurs during this time, we made the assumption that the change in concentration that occurs during the interval Δt is accurately determined by the concentration at the *beginning* of the interval $C(t)$ and we justified this by noting that if the interval of time were sufficiently small, the fractional change in C would be very small.

3. We divided by Δt, which yielded the differential form used in defining a derivative as in Chapter I, and we allowed Δt to go to 0. This allowed us to replace the differential form $\Delta C/\Delta t$ by dC/dt and yielded the differential equation (2.1).

4. We then stated a solution to the differential equation, which we verified by differentiating and trying it in differential equation (2.1).

5. We then imposed on this solution an initial condition, that the concentration at zero time had to be the stated initial concentration of 0.3 mole/liter. The solution was then complete.

The solution of this problem has a form, exponential decay, that occurs frequently in biological problems. Exponential decays occur in those cases where the rate of change of a variable, in this case, concentration, is dependent upon the value of the variable. To say this another way, at any time, the rate at which tracer is washed out is proportional to the amount of tracer in the tank at that time.

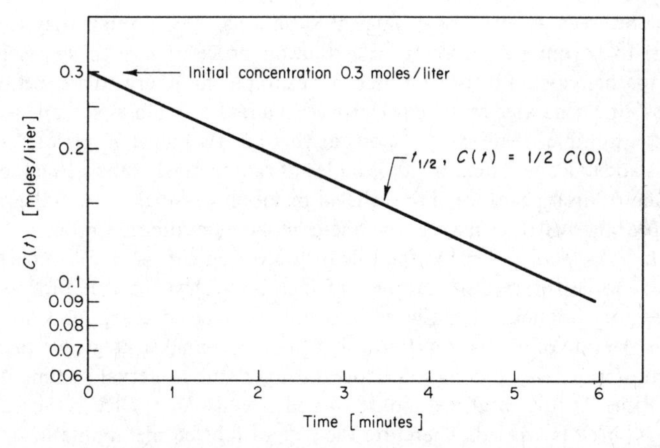

Figure 2-3. The concentration in a one-compartment continuous dilution as a function of time plotted semilogarithmically.

Figure 2-3 shows the solution plotted on semilogarithmic paper, which (as we recall from Chapter I) plots decaying exponentials as straight lines. We construct the plot by noting that C starts from 0.3 mole/liter and decays to one-half its initial value when $(F/V)t = 0.69$ (see Chapter I):

$$t_{1/2} = \frac{0.69V}{F} = \frac{0.69 \times 10 \text{ liters}}{2 \text{ liters/min}} = 3.45 \text{ min}$$

Although basic to the understanding of many dilution processes, the one-compartment dilution rarely arises in this ideal form. Most biological situations are complicated by the existence of two, three, four, or even more compartments in parallel; for example, intracellular volumes versus extracellular volumes versus blood volume, or in the case of organ perfusion, various parts of the organ may act as separate compartments. There are, however, cases in which compartments beyond the first are so small relative to the first compartment that a simple exponential decay is observed over a concentration range of the order of a hundred to one or even more. Thus, simple dilution has been used to characterize the washout of tracers from coronary arteries or whole limbs. A discussion of the resolution of multicompartments will be postponed until a later chapter.

2. One-Compartment Metabolic Turnover

There are a number of biological situations that, though they initially appear to be quite different from a dilution process, nevertheless yield the same mathematical forms. Consider, for example, an intermediate metabolite that is being generated and then destroyed at a rate of R moles/sec, and assume that at any time there exists a pool of this metabolite of Q moles. To this pool is added a small quantity $q(0)$ moles of radioactively labeled metabolite. The rate of disappearance of the labeled metabolite is analogous to the rate of disappearance of the tracer in the one-compartment dilution process. In this case the total pool $Q + q$ has the role of the volume of the one-compartment process, while the rate of turnover of metabolite has the role of flow. We assume for simplicity that the addition of the labeled compound does not change the rate of utilization. Returning to our original strategy, we calculate the amount of labeled metabolite lost during a small interval of time Δt.

In time Δt, the total metabolite turned over is $R\,\Delta t$. Of this the fraction $q(t)/[Q+q(t)]$ is labeled. Therefore the loss of labeled metabolite in Δt is

$$\Delta q = -\frac{q(t)}{Q+q(t)}\,\Delta t\,R$$

$$\text{moles} = \frac{\text{moles}}{\text{moles}}\quad \text{sec}\quad \frac{\text{moles}}{\text{sec}}$$

Since $q(t)$ is small compared to Q, we may make the approximation

$$\Delta q = -\frac{q(t)}{Q}\,R\,\Delta t \tag{2.4}$$

Dividing by Δt and taking the limit as Δt goes to zero, we find a differential equation that is identical in form to Eq. (2.1) of the one-compartment dilution process

$$\frac{dq(t)}{dt} = -\frac{R}{Q}q(t)$$

Its solution is therefore

$$q(t) = q(0)e^{-(R/Q)t} \tag{2.5}$$

which the reader should verify by differentiating and trying the resulting derivative in the differential equation.

Unlike the one-compartment dilution, this process is frequently encountered in its exact form in many metabolic studies.

3. Radioactive Decay

Still another example of the same mathematical form is given by the decay of a radioactive isotope. In this case, the number of atoms that decay per unit time is proportional to the number of remaining radioactive atoms:

$$\Delta N = -KN\,\Delta t$$

$$\frac{dN}{dt} = -KN \tag{2.6}$$

$$N = N(0)e^{-kt}$$

4. X-Ray Absorption

A slightly different example, still yielding the same mathematical form, is given by the absorption of x rays through a homogeneous partially opaque body. In this case, the decay of x-ray intensity is not a function of time but rather a function of penetration distance. The intensity is written as $I(x)$ and we consider a thin slice of the object perpendicular to the direction of incidence of the x rays as shown in Figure 2-4. Entering the left side of the thin slice we have an intensity $I(x)$. Coming out of the thin slice we have $I(x + \Delta x)$ where Δx is the thickness of the slice. The absorption, the difference between $I(x + \Delta x)$ and $I(x)$, is proportional to the intensity, the density of the medium, and its thickness. Thus we may write

$$\Delta I = I(x + \Delta x) - I(x) = -DI(x)\,\Delta x$$

and dividing through by Δx, we take the limit as Δx goes to zero

$$\frac{dI(x)}{dx} = -DI(x) \tag{2.7}$$

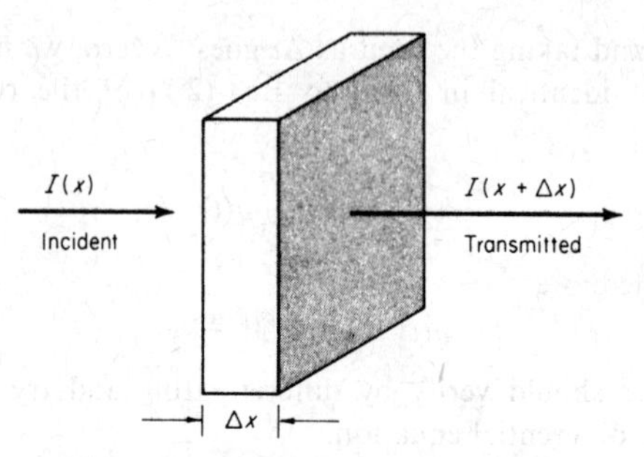

Figure 2-4. Model for x-ray absorption.

The resulting equation has exactly the same form as the one-compartment dilution problem, even though the physical context is quite different. We may write the intensity of the x rays in terms of the incident intensity $I(0)$ at the left edge of a thick absorbing block as

$$I(x) = I(0)e^{-Dx} \tag{2.8}$$

Within the assumption of homogeneous absorption this result is also exact.

5. One Compartment with Constant Injection

Returning to the one-compartment dilution, we consider the same physical situation but with the tracer injected at a constant rate of R moles/sec beginning at time 0. As in the previous case the amount of tracer lost from the tank in a small unit of time Δt will be

$$\Delta Q_{\text{lost}} = -\Delta t\, FC(t)$$

but at the same time tracer is entering the tank at a rate R, so that in the interval Δt the net change in quantity of tracer in the tank is given by

$$\Delta Q_{\text{net}} = -\Delta t\, FC(t) + \Delta t\, R$$

$$\Delta C = \frac{1}{V}(R - FC(t))\,\Delta t$$

Once again divide by Δt

$$\frac{\Delta C}{\Delta t} = \frac{R - FC(t)}{V}$$

and take the limit as Δt goes to 0. Thus we get our second differential equation

$$\frac{dC(t)}{dt} = \frac{R}{V} - \frac{F}{V}C(t) \tag{2.9}$$

$$\frac{\dfrac{\text{moles}}{\text{liter}}}{\text{sec}} = \frac{\dfrac{\text{moles}}{\text{sec}}}{\text{liters}} - \frac{\dfrac{\text{liter}}{\text{sec}}}{\text{liters}} \; \frac{\text{moles}}{\text{liter}}$$

to which the solution is

$$C(t) = \frac{R}{F}(1 - e^{-(F/V)t}) \tag{2.10}$$

which is verified by differentiating and trying it in the differential equation

$$\frac{dC}{dt} = \frac{R}{F}\left(0 + \frac{F}{V}e^{-(F/V)t}\right) = \frac{R}{V}e^{-(F/V)t}$$

But from (2.10) we have that

$$e^{-(F/V)t} = 1 - \frac{F}{R}C(t)$$

Thus

$$\frac{dC}{dt} = \frac{R}{V}\left(1 - \frac{F}{R}C(t)\right) = \frac{R}{V} - \frac{F}{V}C(t)$$

We see that (2.9) is satisfied.

Note, however, that this time the initial concentration is not due to an instantaneous injection but is zero, since at time zero when inflow of tracer starts there is negligible tracer in the tank. The reader should verify that the solution yields zero concentration at zero time. At very long times, the concentration becomes

$$C(\infty) = \frac{R}{F}\left(1 - e^{-(F/V)\infty}\right) = \frac{R}{F} \tag{2.11}$$

which is physically reasonable since at very long times a steady state is established, which is due to a continuous mixing of water entering at rate F and tracer entering at a rate R.

Solutions of this type, which involve a term $(1 - e^{-(F/V)t})$, occur frequently enough to warrant giving this form a name. We will call it an *inverted exponential.* It has the characteristic that it starts at some low value, in this case 0, and approaches at infinite time a constant value, in this case R/F.

If one subtracts the concentration at any time from the terminal concentration R/F, the difference will yield a straight line plot on semilogarithmic paper. (See, however, Chapter X on curve fitting.)

6. The Two-Compartment Series Dilution

Now let us consider a somewhat more complicated problem having two compartments in series. The first compartment will be identical to that described in the first example. The second compartment, however, designated by V_2, will be filled by the effluent from the first compartment and it, in turn, will be a flow-through-type compartment, losing solution at a rate F. The concentration in the first compartment will be identical to that given in the first example, (see Figure 2-5):

$$C_1(t) = C_1(0)e^{-(F/V_1)t}$$

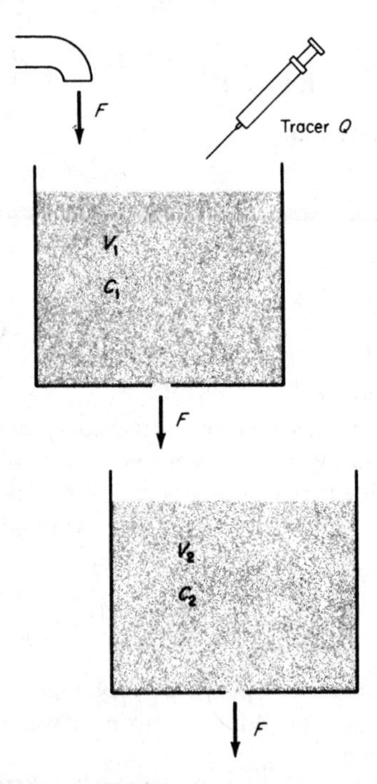

Figure 2-5. **Model for the two-compartment series dilution problem.**

To calculate the concentration in the second compartment as a function of time, we return to our basic strategy of finding the total amount of tracer that enters and leaves the compartment in a small interval of time Δt. The amount leaving will be the rate of flow F multiplied by the concentration in the second compartment C_2 multiplied by the small time interval Δt. The amount entering during that time will be the flow rate F multiplied by the interval of time Δt multiplied by the concentration of the effluent of the first compartment C_1. Thus the net change in the quantity of tracer in the second compartment in a small interval of time is given by

$$\Delta Q_2 = -\Delta t\, C_2(t)F + \Delta t\, C_1(t)F \tag{2.12}$$

and this is converted to a change of concentration by dividing by the volume V_2.

$$\Delta C_2 = \frac{F}{V_2} C_1(t)\,\Delta t - C_2(t)\frac{F}{V_2}\Delta t$$

Once again divide by Δt and take the limit as Δt goes to 0; this yields

$$\frac{dC_2}{dt} = \frac{F}{V_2}[C_1(t) - C_2(t)] = \frac{F}{V_2}[C_1(0)e^{-(F/V_1)t} - C_2(t)] \tag{2.13}$$

to which the solution is

$$C_2(t) = \frac{V_1 C_1(0)}{V_1 - V_2}[e^{-(F/V_1)t} - e^{-(F/V_2)t}] \tag{2.14}$$

which is verified by transposing (2.13) and inserting C_2 and its derivative from (2.14)

$$\begin{aligned}\frac{dC_2}{dt} + \frac{F}{V_2}C_2 &= \frac{V_1 C_1(0)}{V_1 - V_2}\left[\left(-\frac{F}{V_1} + \frac{F}{V_2}\right)e^{-(F/V_1)t} + \left(\frac{F}{V_2} - \frac{F}{V_2}\right)e^{-(F/V_2)t}\right]\\ &= \frac{V_1 C_1(0)}{V_1 - V_2}F\left[\frac{-V_2 + V_1}{V_1 V_2}e^{-(F/V_1)t}\right]\\ &= \frac{FC_1(0)}{V_2}e^{-(F/V_1)t}\end{aligned}$$

We see that (2.13) is satisfied. The reader should check the dimensions of (2.14).

7. Dimensionless Variables

When we encounter a functional relationship as complicated as (2.14), we usually wish to visualize it by plotting $C_2(t)$ as a function of time. In doing so, we are confronted with the fact that V_1, V_2, C_1, and F can all take on wide ranges of values in practical problems, and the labor involved in making plots for wide ranges of four different parameters is prohibitive. We therefore examine the function to see if it is possible to reduce the number of parameters in some simple way, and in the case of (2.14), at least one way is obvious. We note that $V_1C_1(0)$ is simply the initial quantity of tracer in the first compartment. It is quite evident that if we were to double this quantity, we would double $C_2(t)$. Perhaps the appropriate procedure is to plot not $C_2(t)$ but the ratio of $C_2(t)$ to the initial quantity of tracer. The problem, however, would still be fairly complicated, so let us look for other simplifications. We note that the quantities F and t do not appear independently, but always in their product. Therefore, we might guess that we ought to replace the product Ft by a single variable, but note further that the two exponents differ only by the ratio of V_1 to V_2, so perhaps the thing to do is to replace the quantity Ft/V_1 by a variable, which we shall call p, and Ft/V_2 by pK where K is the ratio of the two volumes. Now it is not obvious that this last step is a better thing to do than to just replace the product Ft, but in general, experience indicates that graphical results are simpler in terms of dimensionless quantities, and we note that Ft/V is a dimensionless quantity. It is volume per time F multiplied by time divided by volume, and the ratio K is the ratio of two volumes, so that by proceeding in this way we get simple dimensionless quantities in the exponents. It would be nice if we could express the leading fraction in terms of the same dimensionless quantities. This is easily done by dividing through by V_1 and thus we find

$$\frac{C_2(t)}{C_1(0)} = \frac{1}{1-(1/K)}(e^{-p} - e^{-Kp})$$

where $K = V_1/V_2$. This function, which is a ratio of concentrations and is therefore dimensionless, is expressed in terms of only two dimensionless variables p and K. Thus a problem that appeared to have four independent variables is reduced to one of only two independent, and now dimensionless, variables, so that a plot of C_2 over $C_1(0)$ is easily made (Figure 2-6). We

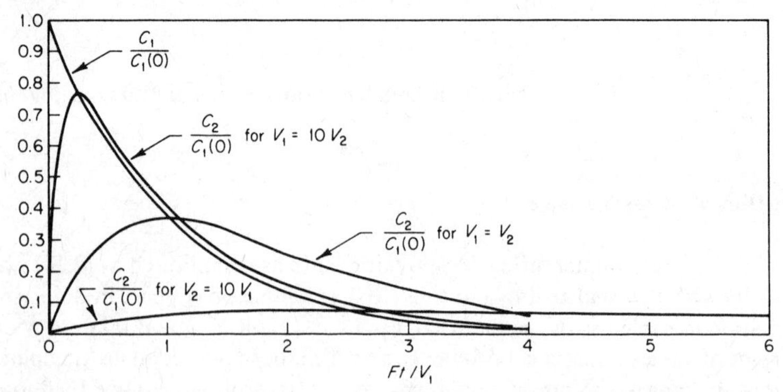

Figure 2-6. The concentration in the second compartment divided by the initial concentration of the first compartment of a two-compartment series dilution. Plotted linearly.

observe from this plot the intuitively satisfying result that if V_1 is large compared to V_2, the concentration in the second compartment follows very closely the concentration in the first compartment, which is not really surprising, and therefore the system will act very much like a one-compartment system. Similarly, if V_2 is very large compared to V_1, although their concentrations will differ by a great deal, nevertheless, as viewed in compartment V_2, it will still appear to be very much like a one-compartment problem. Only in those cases where the compartments are somewhat similar in size is a significant deviation from one-compartment behavior observable. This observation leads us to be a bit suspicious of a plot in the form of Figure 2-6. Perhaps it hides something that we ought to know about the time dependence of C_2. Let us return to Eq. (2.14) and observe that it has a rather peculiar characteristic. If written in terms of initial quantity Q

$$C_2(t) = \frac{Q}{V_1 - V_2}[e^{-(F/V_1)t} - e^{-(F/V_2)t}]$$

for any fixed value of the parameters, the numbers for V_1 and V_2 can be interchanged without changing the result. The interchange V_1 and V_2 simply reverses the order of the terms in the bracket and in the denominator, but since they both reverse, the result is unchanged. It would be nice if we could plot our results in such a way as to make this apparent. It is clear from the previous discussion that the ratio of V_1 to V_2 is an important parameter. But what parameter will remain unchanged when V_1 and V_2 are interchanged? One such parameter would be the sum of V_1 and V_2, so let us try a plot of the concentration in the second compartment divided, not by the concentration in the first compartment, but by the size of the injection Q divided by the volume of the sum of the two compartments. With a bit of algebraic manipulation, we find that if

$$V_{\mathrm{T}} = V_1 + V_2 = V_2(1 + K)$$

then

$$\frac{1}{V_1} = \frac{1 + K}{KV_{\mathrm{T}}} \quad \text{and} \quad \frac{1}{V_2} = \frac{1 + K}{V_{\mathrm{T}}}$$

We let $p' = Ft/V_{\mathrm{T}}$; then

$$\frac{C_2(t)}{Q/V_T} = \frac{K+1}{K-1}[e^{-p'(1+1/K)} - e^{-p'(1+K)}] \tag{2.15}$$

and in this form we note that if K is replaced by its reciprocal (which is equivalent to interchanging V_1 and V_2), the result remains unchanged. Figure 2-7 is a plot of C_2 divided by the initial quantity of tracer in turn divided by the sum of the two volumes. We observe that, except for the curve $K = 1$, each curve represents two different ratios of V_1 to V_2. For example, $K = 10$ can be V_2 equal to 10 times V_1 or V_1 equal to 10 times V_2. We have now displayed the result in a far more significant way, again in terms of carefully chosen dimensionless parameters. This way has the great advantage that as K changes, the shape of the curve changes very little; therefore, unlike our first plot, it is very easy to interpolate to find the shape of the curve for other values of K. It also tells us one other thing that was perhaps not evident. From a plot of concentration versus time in the second compartment, we can infer the ratio of the volumes of the two compartments. We cannot, however, tell which of the compartments is larger. Of course, if we can measure C_1 simultaneously, or if we can measure the actual quantity in either compartment, then we can determine both the ratio of the sizes of the compartments and which is the larger.

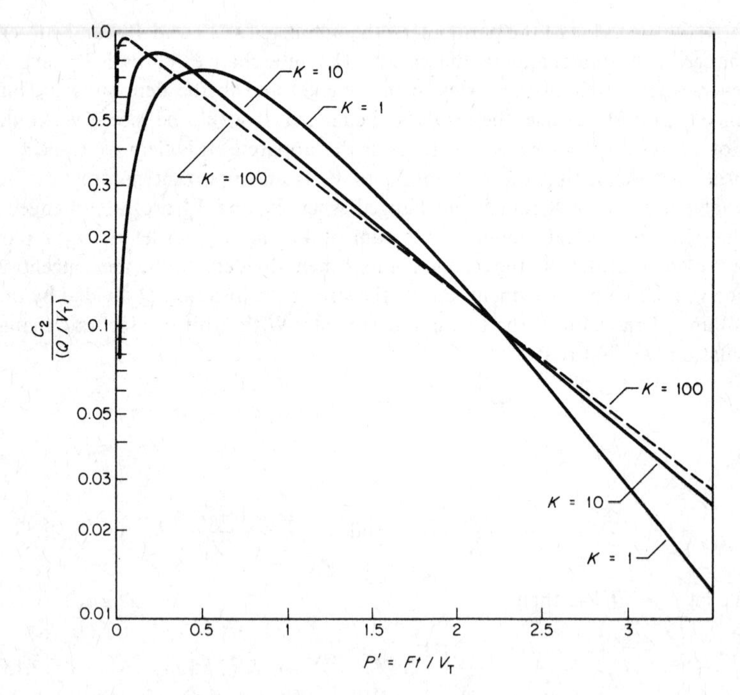

Figure 2-7. A dimensionless plot of the concentration in the second compartment in a two-compartment series dilution plotted semilogarithmically.

8. Mother–Daughter Radioactive Decay

Like the one-compartment dilution problem, the two-compartment series dilution has analogs in radioactive and metabolic processes. For example, many radioactive elements have decay products that are themselves radioactive. Let us consider a mother element with half-life τ_1 and a daughter element with half-life τ_2. Recalling that the half-life is that value of time at which the exponent of e is minus 0.69, we write for the amount of mother element at any time, the equation

$$M(t) = M(0)e^{-0.69(t/\tau_1)} \tag{2.16}$$

Thus the rate at which mother element is lost is given by

$$\frac{dM}{dt} = -M(0)\frac{0.69}{\tau_1}e^{-0.69(t/\tau_1)} = -M(t)\frac{0.69}{\tau_1} \tag{2.17}$$

We assume that for each mother atom decayed, a daughter atom is produced. The daughter atoms in turn are lost in proportion to their own quantity; thus the net rate of change of daughter substance is given by

$$\begin{aligned}\frac{dD(t)}{dt} &= -D(t)\frac{0.69}{\tau_2} + M(t)\frac{0.69}{\tau_1}\\ &= -D(t)\frac{0.69}{\tau_2} + M(0)\frac{0.69}{\tau_1}e^{-0.69(t/\tau_1)}\end{aligned}$$

which is immediately seen to be analogous to Eq. (2.13) for the second compartment of a two-compartment series dilution.

9. A Circular Reaction

Here is another interesting example of a type that occurs frequently in metabolic studies. Figure 2-8 represents a circular reaction. In the actual

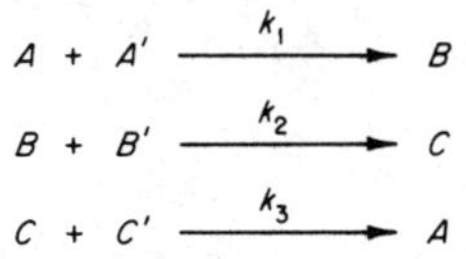

$$A + A' \xrightarrow{k_1} B$$

$$B + B' \xrightarrow{k_2} C$$

$$C + C' \xrightarrow{k_3} A$$

Figure 2-8. Representation of a circular reaction.

biological case, it would usually have inputs and outputs from the circle, but for simplicity we consider only the circle. The compounds A and A' combine according to kinetic principles to produce B at a rate equal to the product of the concentrations of A and A' and a rate constant k_1. Then B reacts with B' to form C, and C reacts with C' to replenish A. In a short interval of time Δt, B is generated according to

$$\Delta B_g = AA'k_1 \, \Delta t$$

but in the same interval of time it is lost according to

$$\Delta B_l = BB'k_2 \, \Delta t$$

Thus the net change in B is

$$\Delta B = \Delta t(AA'k_1 - BB'k_2)$$

Dividing by Δt and taking the limit as $t \to 0$ we get

$$\frac{dB}{dt} = AA'k_1 - BB'k_2$$

Similarly,

$$\frac{dC}{dt} = BB'k_2 - CC'k_3 \quad \text{and} \quad \frac{dA}{dt} = CC'k_3 - AA'k_1$$

Let us solve this system for the steady state, that is, when none of the concentrations are changing with time. Each of the foregoing derivatives is zero. Let us further simplify the notation by letting

$$K_1 = A'k_1, \qquad K_2 = B'k_2, \qquad K_3 = C'k_3$$

Then

$$AK_1 = BK_2, \qquad BK_2 = CK_3, \qquad CK_3 = AK_1$$

Let

$$Q = A + B + C = A + \frac{K_3}{K_2} C + \frac{K_1}{K_3} A = A\left(1 + \frac{K_1}{K_2} + \frac{K_1}{K_3}\right)$$

Then

$$A = \frac{Q}{1 + \dfrac{K_1}{K_2} + \dfrac{K_1}{K_3}}, \qquad B = \frac{K_1}{K_2} A, \qquad C = \frac{K_1}{K_3} A$$

In Chapter VI we will see what happens to a system of this type when one of the rate constants changes.

Problems

1. Repeat the first calculation of this chapter for 2% loss per second.

2. Verify Eq. (2.5).

3. Suppose some substance diffuses out of a compartment at a rate proportional to the difference in concentration of the substance within the compartment and the concentration in the surrounding environment, which is assumed constant. Let K be the diffusion constant, C_E the concentration in the environment, C_C the concentration in the compartment, V the volume of the compartment, and $Q(0)$ the initial quantity of the substance in the compartment. Find the differential equation that describes C_C as a function of time.

4. In Problem 3, assume also that the substance is pumped into the compartment at a rate of R moles/sec. Find the differential equation for $C_C(t)$.

5. In a one-compartment dilution the following concentrations were observed as a function of time:

t [sec]	C [moles/liter]
0	0.024
1	0.011
2	0.0048
3	0.0024
4	0.0010

   If $Q(0) = 0.1$ mole, find F and V.

6. In a radioactive decay process half the atoms have decayed in one day. What fraction is left after 10 hours?

7. In the two-compartment dilution suppose a third compartment were added whose input is the output of compartment two. Write the differential equation that describes the concentration of the third compartment.

8. Substance A is produced at a rate of R moles/sec. It is destroyed at a rate of K moles/sec per mole of A. Write the differential equation describing the amount of A as a function of time starting with a zero quantity of A. When the system reaches equilibrium, how much of A is present?

9. In Problem 8 assume that when A is destroyed it is converted into B. However, B is converted back into A at a rate BK'. Write the equations describing A and B as functions of time starting from both A and B zero.

10. Sketch on semi-logarithmic graph paper the curves given by y_1 and y_2

x	y_1	y_2	y_3	y_4
0.0	2.00	1.00	0.00	0.00
0.5	1.21	0.37	0.79	0.63
1.0	0.74	0.13	1.26	0.87
1.5	0.45	0.049	1.55	0.95
2.0	0.27	0.018	1.73	0.98

Write equations which approximately fit the two curves y_1 and y_2.

Now replot on rectangular paper all four curves. By comparing y_1 to y_3 and y_2 to y_4, write equations which fit y_3 and y_4.

III

Guessing the Solution of a Differential Equation

1. Introduction: Quadratic Equations

In this chapter we need to be able to solve quadratic algebraic equations of the form

$$\alpha^2 + B\alpha + D = 0 \tag{3.1}$$

The reader will recall that a quadratic equation has two solutions, which are usually different from each other, although in some cases the two solutions will be equal, and that they are given by

$$\alpha_1 = \tfrac{1}{2}(-B + \sqrt{B^2 - 4D}) \tag{3.2}$$

$$\alpha_2 = \tfrac{1}{2}(-B - \sqrt{B^2 - 4D}) \tag{3.3}$$

which we verify by calculating the square of α_1

$$\begin{aligned}\alpha_1{}^2 &= \tfrac{1}{4}[B^2 - 2B\sqrt{B^2 - 4D} + (B^2 - 4D)] \\ &= \tfrac{1}{2}(B^2 - 2D - B\sqrt{B^2 - 4D})\end{aligned}$$

and adding $B\alpha_1$ to $\alpha_1{}^2$.

$$\begin{aligned}\alpha_1{}^2 + B\alpha_1 &= \tfrac{1}{2}B^2 - D - \tfrac{1}{2}B\sqrt{B^2 - 4D} + \tfrac{1}{2}B(-B + \sqrt{B^2 - 4D}) \\ &= -D\end{aligned}$$

Thus α_1 is a solution of (3.1). (See Problems 1 and 2 at the end of this chapter.)

2. More about Exponential Functions

The reader has no doubt observed that a great number of differential equations have as solutions terms that contain exponential functions of time or exponential functions of distance. In attempting to find solutions to these differential equations, one should think first of the physical problem. If it is such that the dependent variable decreases in time toward a constant value or zero, a decaying exponential should be tried. If it is such that it increases to an upper limit, an inverted exponential can be tried. If the dependent variable starts from one value A and goes at very long times to another stable value B, one constructs as a potential solution a function of the following form, which one can easily see at time zero and time infinity takes on the steady-state values:

$$C = (A - B)e^{-\alpha t} + B \tag{3.4}$$

At $t = 0$,

$$e^{-\alpha t} = 1, \qquad C = A - B + B = A$$

at $t = \infty$,

$$e^{-\alpha t} = 0, \qquad C = B$$

In this chapter we encounter expressions of the form

$$Ae^{\alpha_1 t} + Be^{\alpha_2 t} = 0 \tag{3.5}$$

where A and B are constants whose values must be determined. *If the constants α_1 and α_2 are different, the two constants A and B must both be equal to* 0. We can see this by solving for A in terms of B.

$$A = -Be^{(\alpha_2 - \alpha_1)t}$$

Note the exponential term on the right side of this equation. Its value will change depending on the value of t. Therefore, if A and B have fixed nonzero values, it is not possible for the foregoing equation to be true for more than one value of t. Therefore, we conclude, if the original statement (3.5) was true for more than one value of t, the two constants A and B are both zero.

Furthermore, if A, B, and C are not all zero and

$$Ae^{\alpha_1 t} + Be^{\alpha_2 t} = Ce^{\alpha_3 t} \tag{3 6}$$

and α_1 is different from α_2, then either

$$A = C \quad \text{and} \quad \alpha_1 = \alpha_3 \quad \text{and} \quad B = 0$$

or

$$B = C \quad \text{and} \quad \alpha_2 = \alpha_3 \quad \text{and} \quad A = 0$$

We can prove this by multiplying through by $e^{-\alpha_3 t}$ and differentiating, which yields

$$A(\alpha_1 - \alpha_3)e^{(\alpha_1 - \alpha_3)t} + B(\alpha_2 - \alpha_3)e^{(\alpha_2 - \alpha_3)t} = 0$$

to which we apply the previous rule, so that

$$A(\alpha_1 - \alpha_3) = 0, \qquad B(\alpha_2 - \alpha_3) = 0$$

from which the foregoing result follows. (Do Problems 3 and 4 at the end of this chapter.)

3. Second Derivatives

As we discussed previously, the derivative of a function is a new function. Thus it, too, can have a derivative. We recall that if $y = f(x)$, the derivative of y is the limit as Δx goes to zero of the difference in the value of f at two points separated by Δx, divided by Δx.

$$\frac{dy}{dx} = \operatorname*{limit}_{\Delta x \to 0} \frac{f(x + \Delta x) - f(x)}{\Delta x}$$

Let us call this derivative $p(x)$. Thus

$$p(x) = \operatorname*{limit}_{\Delta x \to 0} \frac{f(x + \Delta x) - f(x)}{\Delta x}$$

But we can also say

$$p(x + \Delta x) = \frac{dy(x + \Delta x)}{dx} = \operatorname*{limit}_{\Delta x \to 0} \frac{f(x + 2\,\Delta x) - f(x + \Delta x)}{\Delta x}$$

and define the derivative of p as

$$\frac{dp}{dx} = \operatorname*{limit}_{\Delta x \to 0} \frac{p(x + \Delta x) - p(x)}{\Delta x} = \operatorname*{limit}_{\Delta x \to 0} \frac{1}{\Delta x}\left(\frac{dy(x + \Delta x)}{dx} - \frac{dy(x)}{dx}\right)$$

$$= \operatorname*{limit}_{\Delta x \to 0} \frac{f(x + 2\,\Delta x) + f(x) - 2f(x + \Delta x)}{(\Delta x)^2} = \frac{d^2y}{dx^2}$$

The quantity on the right is called the second derivative of y with respect to x. Just as the derivative was the rate at which y changed relative to x, the second derivative is the rate at which dy/dx changes relative to x.

Let us calculate some second derivatives.

(a) $y = Ax^2, \qquad \dfrac{dy}{dx} = 2Ax, \qquad \dfrac{d^2y}{dx^2} = 2A$

(b) $y = Ae^{-\alpha t}, \qquad \dfrac{dy}{dt} = -\alpha Ae^{-\alpha t}, \qquad \dfrac{d^2y}{dt^2} = \alpha^2 Ae^{-\alpha t}$

4. Solutions of Simple Differential Equations

The possible solutions of a number of simple differential equations are listed in Table III-1. Each solution contains one or more constants A_1, A_2, and so on, which, from the context of the differential equation, are arbitrary. Most of this chapter is devoted to explaining how one adjusts the values of these constants to fit the given accessory conditions of the physical problem. These solutions are tabulated without proof. Some proofs will be derived within the chapter, others will be left as exercises for the reader.

In each case, the proof consists of differentiating the given solution and demonstrating that it satisfies the differential equation. The letter C represents the unknown or dependent variable, and t is the independent variable. The constants K_1, K_2, and so on, are given by the structure of the problem.

In no circumstances should the solutions in Table III-1 be memorized. In fact, most of them will not appear again in this book. The reader should note some of their properties, however. Note first that the number of arbitrary constants always corresponds to the order of the highest derivative that appears in the equation. Thus Eqs. (1), (2), and (3) contain only a single arbitrary constant, whereas the others, which are of second order, contain two such constants. Notice also that Eqs. (1), (2), (4), and (5), which are characterized by the fact that the unknown C appears only once in the equation, have particularly simple solutions: powers of t multiplied by or added to constants.

Equations (9) and (10) will appear quite often in the remainder of this book. Notice that they differ by the addition of a constant in Eq. (10) and that the value of this constant is such that when it is multiplied by C, it yields the right-hand side of the equation. It is suggested that the reader work Problems 5–8 at the end of this chapter before proceeding.

5. Linear and Nonlinear Equations

Most of the differential equations encountered in this book can be solved by standard methods, some of which are outlined here, and others of which can be found in conventional textbooks on elementary differential equations. However, the reader should endeavor to develop an intuitive approach to the solution of differential equations based on an understanding of the physical nature of the problem. This intuitive approach has a number of virtues: It is often faster than the conventional methods; it fosters the ability to recognize when a solution is reasonable; most important, it develops in the reader a far deeper understanding of the nature of the process he is studying than he would obtain from a "turning the crank" type of solution.

Table III-1

Solutions to Simple Differential Equations

1.	$\dfrac{dC}{dt} = 0$	$C = A_1$
2.	$\dfrac{dC}{dt} = K_1$	$C = K_1 t + A_1$
3.	$\dfrac{dC}{dt} = K_1 C$	$C = A_1 e^{+K_1 t}$
4.	$\dfrac{d^2C}{dt^2} = 0$	$C = A_1 t + A_2$
5.	$\dfrac{d^2C}{dt^2} = K_1$	$C = \dfrac{K_1 t^2}{2} + A_1 t + A_2$
6.	$\dfrac{d^2C}{dt^2} = K_1 C, \quad K_1 > 0,$	$C = A_1 e^{t\sqrt{K_1}} + A_2 e^{-t\sqrt{K_1}}$
7.	$\dfrac{d^2C}{dt^2} + K_1 \dfrac{dC}{dt} = 0$	$C = \dfrac{A_1}{K_1}(1 - e^{-K_1 t}) + A_2$
8.	$\dfrac{d^2C}{dt^2} + K_1 \dfrac{dC}{dt} = K_2$	$C = \dfrac{K_2}{K_1} t + \dfrac{A_1}{K_1}(1 - e^{-K_1 t}) + A_2$
9.	$\dfrac{d^2C}{dt^2} + K_1 \dfrac{dC}{dt} + K_2 C = 0$	$C = A_1 e^{\alpha_1 t} + A_2 e^{\alpha_2 t}$
	$K_1{}^2 > 4K_2$	$\alpha_1 = -\frac{1}{2}(K_1 + \sqrt{K_1{}^2 - 4K_2})$
	$K_1 \neq 0$	$\alpha_2 = -\frac{1}{2}(K_1 - \sqrt{K_1{}^2 - 4K_2})$
	$K_2 \neq 0$	
	If $K_1{}^2 = 4K_2$	$C = A_1 e^{-K_1 t/2} + A_2 t e^{-K_1 t/2}$
10.	$\dfrac{d^2C}{dt^2} + K_1 \dfrac{dC}{dt} + K_2 C = K_3$	$C = \dfrac{K_3}{K_2} + A_1 e^{\alpha_1 t} + A_2 e^{\alpha_2 t}$
	$K_1{}^2 > 4K_2$	
	$K_1 \neq 0$	
	$K_2 \neq 0$	
	If $K_1{}^2 = 4K_2$	$C = \dfrac{K_3}{K_2} + A_1 e^{-K_1 t/2} + A_2 t e^{-K_2 t/2}$

We refer in the chapter title to guessing the solution of a differential equation. We chose this title in the spirit of a challenging game that the reader should learn to play. But in fact, we will not usually guess solutions (although skilled mathematicians often do); rather, we shall learn to recognize certain forms of differential equations, in particular, linear differential equations, and we will formulate a method of constructing trial solutions. These trial solutions will contain constants whose values may not be obvious from or determined by the differential equation. However, by considering how the solution must behave for various values of the independent variable, we will quite often be able to assign the correct values to these constants. In those cases in which we are not able to determine the constants in this manner, we differentiate the trial solutions and find what values must be assigned to the constants for the trial solution to fit the given differential equation and its accessory conditions.

All of the differential equations thus far encountered in this book are of a type known as linear. A linear differential equation is one in which the dependent variable or unknown appears only in the first power. Thus, the following differential equations are linear:

$$\frac{d^3C(t)}{dt^3} + 3\frac{d^2C(t)}{dt^2} + t^2C(t) = 0,$$

$$\frac{d^2C(t)}{dt^2} + C(t) = 0$$

In contrast, these equations are nonlinear:

$$\frac{d^2C(t)}{dt^2} + C^2(t) = 0$$

$$\frac{d\sqrt{C(t)}}{dt} + C(t) = 0$$

6. Homogeneous and Inhomogeneous Equations

In addition to distinguishing between linear and nonlinear differential equations, it is useful to distinguish between homogeneous and inhomogeneous equations. When the dependent variable appears exactly once in every term of the equation, the equation is said to be homogeneous. (Most books on differential equations use a more general definition of homogeneous, of which this is a special case.) Thus, the following equations are homogeneous:

$$\frac{d^2C(t)}{dt^2} + t\frac{dC(t)}{dt} + 4C(t) = 0$$

$$\sin \omega t \frac{dC(t)}{dt} + C(t) \cos \omega t = 0$$

If there are terms in the equation that do not contain the dependent variable, the equations, like the following examples, are called inhomogeneous:

$$\frac{dC(t)}{dt} + C(t) = 4,$$

$$\frac{d^2C(t)}{dt^2} + \frac{dC(t)}{dt} + t = 0$$

In addition to the foregoing definitions, we need one more. We refer to the order of a differential equation, which signifies the highest-order derivative that appears in the equation. Thus the equation immediately preceding is a second-order linear inhomogeneous equation.

7. The Forms of Solutions

Equations that are linear and homogeneous and have *constant coefficients* have a particularly simple kind of solution. The solution can always be expressed as a sum of exponentials, each of which may be multiplied by an arbitrary constant. The number of such exponential terms required for a solution is the same as the order of the differential equation. Thus the differential equation

$$\frac{d^2C(t)}{dt^2} + B\frac{dC(t)}{dt} + DC(t) = 0 \tag{3.7}$$

which is second-order, linear, and homogeneous, has the solution

$$C(t) = A_1 e^{\alpha_1 t} + A_2 e^{\alpha_2 t} \tag{3.8}$$

where the constants A_1 and A_2 are arbitrary and α_1 and α_2 are solutions of the quadratic algebraic equation

$$\alpha^2 + B\alpha + D = 0 \tag{3.9}$$

We can verify that (3.8) is a solution by differentiating the proposed solution and noting that when the derivatives are inserted into the differential equation, the equation is indeed satisfied.

$$\frac{dC}{dt} = A_1\alpha_1 e^{\alpha_1 t} + A_2 \alpha_2 e^{\alpha_2 t}$$

$$\frac{d^2C}{dt^2} = A_1\alpha_1{}^2 e^{\alpha_1 t} + A_2 \alpha_2{}^2 e^{\alpha_2 t}$$

Therefore

$$\frac{d^2C}{dt^2} + B\frac{dC}{dt} + DC = A_1({\alpha_1}^2 + \alpha_1 B + D)e^{\alpha_1 t} + A_2({\alpha_2}^2 + \alpha_2 B + D)e^{\alpha_2 t}$$

Since α_1 and α_2 are solutions of (3.9), the right-hand side of the equation above is zero and Eq. (3.7) is satisfied.

In special cases equations of this type may have other kinds of solutions. For example, if

$$B^2 = 4D$$

the equation will have a solution of the type

$$C(t) = Ate^{\alpha t}$$

and in the next chapter we will see how to find these solutions. For the moment, however, we restrict our attention to exponentials or, as in the following case, exponential, plus a constant term.

Now let us consider a similar differential equation but with the addition of an inhomogeneous term:

$$\frac{d^2C}{dt^2} + B\frac{dC}{dt} + DC = 4 \tag{3.10}$$

The solution to this equation is given by

$$C(t) = A_1 e^{\alpha_1 t} + A_2 e^{\alpha_2 t} + \frac{4}{D} \tag{3.11}$$

which the reader should verify by trial, as is done in the homogeneous case. We note that the solution of the inhomogeneous equation is the same as that of the homogeneous equation with the exception that we have added the value 4/D, which is required to match the inhomogeneous term on the right side of Eq. (3.10). The solution of an inhomogeneous linear equation will always consist of the solution to the homogeneous equation plus the addition of a constant or a function of the independent variable. Cases in which the solution to the inhomogeneous equation can be produced by adding a constant to the solution of the homogeneous equation occur frequently enough to warrant our taking special note of this case.

The arbitrary constants A_1 and A_2 that appear in the foregoing solutions cannot be determined from the differential equation. Their values must be determined by some accessory condition of the problem. Thus, in the solution of the one-compartment dilution problem, the arbitrary constant was chosen to be the initial concentration at time zero. In the two-compartment series dilution problem, the initial concentration of the first compartment is one accessory condition, while the initial concentration of the second compartment (which is zero) is another.

8. The Two-Compartment Series Dilution

Let us use the two-compartment series dilution as an example of the solution of an inhomogeneous linear differential equation. We recall from the preceding chapter that the input to the second compartment consists of a steady fluid flow but with an exponentially decaying concentration. Noting that the initial concentration of the second compartment is 0 and that at very long times it must return to 0, we can sketch the concentration in the second compartment roughly as shown in Figure 3-1. The differential equation that describes this concentration is

$$\frac{dC_2}{dt} = \frac{F}{V_2}\left(C_1(t) - C_2(t)\right) = \frac{F}{V_2}\left(C_1(0)e^{-(F/V_1)t} - C_2(t)\right) \tag{3.12}$$

which we derived in Chapter II. We observe that it is linear but inhomogeneous. We would therefore suspect solutions of the exponential type even without reference to Table III-1. However, the solution cannot be a single exponential, since a single exponential cannot both start at zero, which the concentration in the second compartment does, and return to zero at large values of time, which physically the concentration must do. Perhaps it can

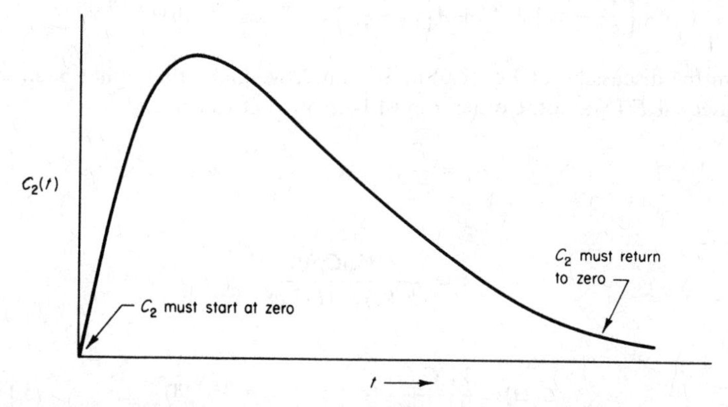

Figure 3-1. Sketch of what we expect the concentration in the second compartment of a two-compartment series dilution to be like for instantaneous input to the first compartment. Note that the function must start at zero concentration and must return to zero concentration at long times, which suggests a double exponential-type solution.

be the sum of two exponentials. Let us, therefore, try a solution of the form

$$C_2(t) = A_1 e^{-\alpha_1 t} + A_2 e^{-\alpha_2 t} \tag{3.13}$$

and by differentiating it, see whether or not we can make it satisfy the differential equation and the accessory conditions that the concentration must be zero for both zero time and infinite time.

Now, if the solution is to be the sum of two exponential terms, there are two conditions that must be true. We know that both exponentials of the trial solution must have negative exponents, since a positive exponent would imply a concentration that becomes infinite at infinite time. We further know that the constants A_1 and A_2 must be equal to each other but of opposite signs, so that their sum will be zero at zero time.† Thus, if a double exponential is to be the solution to Eq. (3.12), it must be of the form

$$C_2(t) = A(e^{-\alpha_1 t} - e^{-\alpha_2 t}) \tag{3.14}$$

Let us try this solution by differentiating it and seeing if it can be made to fit.

$$\frac{dC_2}{dt} = A(-\alpha_1 e^{-\alpha_1 t} + \alpha_2 e^{-\alpha_2 t})$$

This must be equal to the right-hand side of (3.12). Therefore,

$$A\alpha_2 e^{-\alpha_2 t} - A\alpha_1 e^{-\alpha_1 t} = \frac{F}{V_2} C_1(0) e^{-(F/V_1)t} - \frac{F}{V_2} A e^{-\alpha_1 t} + \frac{F}{V_2} A e^{-\alpha_2 t}$$

or

$$A\left(\frac{F}{V_2} - \alpha_1\right) e^{-\alpha_1 t} + A\left(\alpha_2 - \frac{F}{V_2}\right) e^{-\alpha_2 t} = \frac{F}{V_2} C_1(0) e^{-(F/V_1)t}$$

From the discussion of Eq. (3.6) in Section 2, we know that either α_1 or α_2 must equal F/V_2 and the other α must be F/V_1. Let us choose

$$\alpha_1 = \frac{F}{V_2}, \qquad \alpha_2 = \frac{F}{V_1}$$

Thus

$$A = \frac{(F/V_2)C_1(0)}{(F/V_1) - (F/V_2)}$$

and

$$C_2(t) = \frac{V_1 C_1(0)}{V_1 - V_2} (e^{-(F/V_1)t} - e^{-(F/V_2)t}) \tag{3.15}$$

† The philosophy here might best be described as "... when you have eliminated the impossible, whatever remains, ... must be the truth." Attributed to S. H. by J. W. (10).

9. One Compartment with a Long Input

In Chapter II we discussed the one-compartment dilution problem for the case in which the tracer was injected instantaneously at time zero and for another case in which the tracer was injected at a constant rate R. Now let us consider a problem that has practical application in the measurement of blood flow by dilution in which one attempts an instantaneous injection but in fact achieves an injection of tracer at a fixed rate for some short but nonzero time.

From our previous study of the single-compartment dilution problem, we know that the differential equation (2.9) to be satisfied is

$$V\frac{dC}{dt} = R(t) - FC \tag{3.16}$$

where the input function $R(t)$ now is a constant rate of input R for, say, 10 sec, followed by no additional input. Initial concentration of tracer is of course zero. While the injection of tracer is going on, the concentration of tracer in the compartment will increase, reaching some value $C(10)$ at time 10 sec. Thereafter, the concentration of tracer in the compartment must decay toward zero. It seems entirely reasonable that this decay would be a simple decreasing exponential beginning at time 10 sec. Let us therefore try a solution of this form, assuming that in some way we will be able to determine $C(10)$, the concentration at 10 sec. Our trial solution therefore is

$$C(t) = C(10)e^{-(F/V)(t-10)}, \qquad t > 10 \text{ sec} \tag{3.17}$$

The reader can immediately verify that this satisfies (3.16) for $R(t) = 0$, the condition after 10 sec. In order to determine $C(10)$ let us look at a related problem. Instead of an injection that lasts only 10 sec, let us consider a constant rate of injection beginning at time zero and lasting forever. Since the rate of flow is F and the rate of injection of tracer is R, it is clear that at infinite time the concentration will approach the value R/F. Since it starts at zero, let us try the exponential form, which we know starts at zero and approaches at infinite time a fixed value R/F. Our trial solution is therefore

$$C(t) = \frac{R}{F}(1 - e^{-\beta t}) \tag{3.18}$$

which when differentiated is found to satisfy Eq. (3.16) if the constant β is chosen to be equal to F/V. Thus we have solved the problem of the concentration during a 10-sec injection by considering not the original problem, but the related problem of a constant injection beginning at time zero and lasting for all time thereafter. $C(10)$ is now known to be

$$C(10) = \frac{R}{F}(1 - e^{-10(F/V)}) \tag{3.19}$$

(F in volume per second) and the complete solution of our problem is given by

$$C(t) = \frac{R}{F}(1 - e^{-(F/V)t}), \qquad t < 10 \text{ sec} \tag{3.20}$$

$$= \frac{R}{F} e^{-(F/V)(t-10)}(1 - e^{-10(F/V)}), \qquad t > 10 \text{ sec} \tag{3.21}$$

Now let us try a more complicated example.

10. Diffusion between Compartments

Two compartments are separated by a barrier through which solute can diffuse. The rate of diffusion of solute from one compartment to the other is proportional to the difference in concentration of the two compartments C_1 and C_2, and will be from the higher concentration to the lower. Thus in a short interval of time Δt the amount of solute that will cross the barrier will be

$$\Delta Q = \Delta t \, K(C_1 - C_2)$$

where K is a constant depending on the nature of the barrier and solute and the geometry of the barrier (as well as other things). Thus,

$$\frac{\Delta C_1}{\Delta t} = \frac{1}{V_1}\frac{\Delta Q_1}{\Delta t} = \frac{K}{V_1}(C_2 - C_1),$$

$$\frac{\Delta C_2}{\Delta t} = \frac{1}{V_2}\frac{\Delta Q_2}{\Delta t} = \frac{K}{V_2}(C_1 - C_2)$$

where V_1 and V_2 are the volumes of the two compartments, which are assumed constant. In the limit as Δt goes to zero

$$V_1 \frac{dC_1}{dt} = K(C_2 - C_1) \tag{3.22a}$$

and

$$V_2 \frac{dC_2}{dt} = K(C_1 - C_2) = -V_1 \frac{dC_1}{dt} \tag{3.22b}$$

Let the initial concentration in the two compartments be $C_1(0)$ and $C_2(0)$. Thus in the initial state, the total amount of solute present is

$$Q_T = V_1 C_1(0) + V_2 C_2(0)$$

After sufficient time an equilibrium condition is reached in which the concentrations in the two compartments are equal. Thus the solute is now distributed throughout a volume V_1 plus V_2. Therefore the concentration at equilibrium is given by

$$C(\infty) = \frac{V_1 C_1(0) + V_2 C_2(0)}{V_1 + V_2} \tag{3.23}$$

Note that this result is obtained independently of the nature of the diffusion process. One needs only to know that an equilibrium is finally reached.

We can now draw a rough sketch (Figure 3-2) of the concentrations as

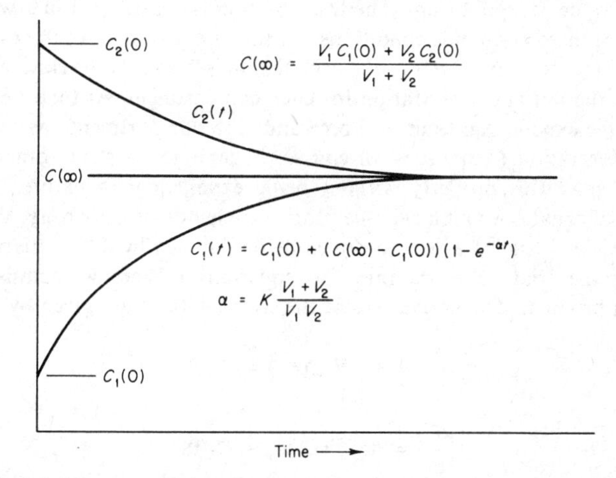

Figure 3-2. The concentrations in two compartments connected by a diffusing barrier as a function of time. Note that if the initial concentrations and volumes are known, the concentration at infinite time is also known, and that the concentrations in the two compartments must approach this value exponentially.

functions of time. Let us attempt to find analytically the time course of the concentrations as they approach equilibrium. We know that the greatest difference in concentration between the two compartments exists at time zero and that this difference will grow progressively smaller as equilibrium is approached. Therefore, the curves of concentration versus time will initially approach each other rapidly but then reduce their rate of convergence as the concentrations become more nearly equal. It is therefore reasonable to assume that these curves will have an exponential-like behavior (decaying or inverted) in which compartment one will start from $C_1(0)$ and go to the final value $C(\infty)$ while compartment two will start from $C_2(0)$ and go to $C(\infty)$. One might therefore guess at trial solutions for C_1 and C_2 as a function of time in the form

$$C_1 = C_1(0) + (1 - e^{-\alpha t})(C(\infty) - C_1(0)) \tag{3.24}$$

$$C_2 = C_2(0) + (1 - e^{-\beta t})(C(\infty) - C_2(0)) \tag{3.25}$$

where the constants α and β will be determined by differentiating the trial solutions and inserting them into the differential equations.

We have referred to this process as guessing a solution. It is not, however, a guess in the sense of guessing the outcome of a turn of a roulette wheel, but rather an educated guess based on the fact that the concentration in each compartment must start from a known initial value and must reach a known terminal value at equilibrium. The trial solution is constructed in a way that guarantees these end-point conditions. At time t equal to zero, the exponentials are equal to 1, and 1 minus the exponential is 0, so that the concentration is simply the initial concentration for each compartment. At time t equal to infinity, the exponential terms are zero and each compartment has the terminal concentration $C(\infty)$. It is, of course, possible to construct other forms that will have this property without being exponential in nature, but the differential equations are linear; therefore, exponential is a good try. Whether or not this will lead to a solution must be determined by differentiating and inserting the trial solutions into the equations of the two-compartment diffusion problem. The derivatives of the trial solutions are given by

$$\frac{dC_1}{dt} = \alpha e^{-\alpha t}(C(\infty) - C_1(0)) \tag{3.26}$$

$$\frac{dC_2}{dt} = \beta e^{-\beta t}(C(\infty) - C_2(0)) \tag{3.27}$$

and from (3.24) and (3.25) the quantity C_2 minus C_1 in Eq. (3.22) is found to be

$$C_2(t) - C_1(t) = C(\infty)(e^{-\alpha t} - e^{-\beta t}) + C_2(0)e^{-\beta t} - C_1(0)e^{-\alpha t} \tag{3.28}$$

These expressions are now inserted into the differential equations (3.22) and one immediately observes that the terms on the right sides contain $C_2 - C_1$, which consists of two different exponential forms, whereas the derivatives on the left each contain only one. The only way of equating them is to choose α equal to β (see Section 2). When this is done

$$C_2(t) - C_1(t) = e^{-\alpha t}(C_2(0) - C_1(0)) \tag{3.29}$$

Therefore,

$$V_1 \frac{dC_1}{dt} = -V_2 \frac{dC_2}{dt} = Ke^{-\alpha t}(C_2(0) - C_1(0))$$

But from (3.26) (multiplying both sides by V_1), we have

$$V_1 \frac{dC_1}{dt} = \alpha V_1 e^{-\alpha t}(C(\infty) - C_1(0))$$

Therefore,

$$\alpha = \frac{K}{V_1} \frac{C_2(0) - C_1(0)}{C(\infty) - C_1(0)} \tag{3.30}$$

which can be simplified by observing from (3.23) that

$$C(\infty) - C_1(0) = \frac{V_1 C_1(0) + V_2 C_2(0) - (V_1 + V_2)C_1(0)}{V_1 + V_2} = \frac{V_2(C_2(0) - C_1(0))}{V_1 + V_2} \tag{3.31}$$

from which we find

$$\alpha = \frac{K(V_1 + V_2)}{V_1 V_2}$$

With this value of α both differential equations are satisfied. Since the form of the solution was designed to ensure that the initial conditions $C_1(0)$ and $C_2(0)$ and the final condition $C(\infty)$ would be satisfied, the problem is solved. The solutions (3.24) and (3.25) can be put into a nicer form by replacing $C(\infty) - C_1(0)$ by its value from (3.31) and $C(\infty) - C_2(0)$ by the analogous form

$$C_1(t) = C_1(0) + (1 - e^{-\alpha t})(C_2(0) - C_1(0)) \frac{V_2}{V_1 + V_2} \tag{3.32}$$

$$C_2(t) = C_2(0) + (1 - e^{-\alpha t})(C_1(0) - C_2(0)) \frac{V_1}{V_1 + V_2} \tag{3.33}$$

One might reasonably ask, were there not other forms, having properties similar to that of the exponential, that might also have been used as trial solutions? The answer is, of course, yes. There are other forms that would have had approximately the same shape. However, since the equations are linear, we knew the exponential form was likely to be the solution.

One of the things we should notice about linear differential equations is that if the inputs to the system are doubled, each and every dependent variable is also doubled. Thus, if we know the concentration, for example, in some compartment of a dilution problem or a diffusion problem as a function of time, when 1 gm of the tracer substance has been injected, we can infer the same time dependence with all of the dependent variables doubled in value for a 2-gm injection. This property allows us to generalize solutions that we have obtained for specific cases, whether these solutions are obtained analytically or empirically, as long as we know that we are dealing with a linear system.

11. A Simple Kinetic Reaction

In general, chemical reactions depend upon concentrations of the reactants. In a particularly simple case, one in which a single molecule of each of the reactants combines to form a product, the rate is usually proportional to the product of the concentrations of the reactants. Thus, if

$$A + B \longrightarrow P$$

the rate at which product is produced is given by

$$\frac{dP}{dt} = K|A||B| \tag{3.34}$$

(The vertical bars denote concentrations.) If this reaction were being observed in a test tube, one would have combined initial concentrations of the substances A and B and observed the rate of production of P. As P is produced, A and B vanish until one of them is exhausted, at which point the reaction stops. Therefore, the concentrations at any time are

$$|A| = \frac{1}{V}(A_{\text{in}} - P), \qquad |B| = \frac{1}{V}(B_{\text{in}} - P)$$

where V is the volume of solution and A_{in} and B_{in} are initial amounts. A complete description of the process is given by

$$\frac{dP}{dt} = \frac{K}{V^2}(A_{\text{in}} - P)(B_{\text{in}} - P)$$

With the techniques we have so far we cannot solve this equation in the general case. However, let us consider a somewhat simplified case in which the initial quantity of A is much greater than the initial quantity of B. Thus, this reaction proceeds until B is exhausted and we can sketch A, B, and P as functions of time roughly as in Figure 3-3. Since the quantity of A changes only slightly relative to its initial size, we can make the approximation that

$$A \approx A_{\text{in}}$$

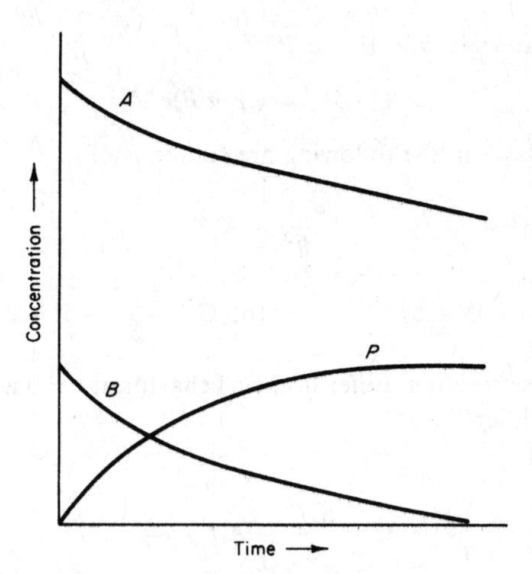

Figure 3-3. The concentrations of substances that combine, according to a simple kinetic principle, to produce a product P.

and therefore

$$\frac{dP}{dt} = \frac{K}{V^2} A_{\text{in}}(B_{\text{in}} - P) \tag{3.35}$$

This is now in the simple form that we recognize as a linear inhomogeneous differential equation in P similar to that encountered in (3.16). It seems reasonable, then, to use as a trial solution an inverted exponential, which reaches the value B_{in} at $t(\infty)$. The trial solution is

$$P = B_{\text{in}}(1 - e^{-\alpha t}) \tag{3.36}$$

When this is differentiated and tried in Eq. (3.35), it is found that

$$\alpha = \frac{KA_{\text{in}}}{V^2} \tag{3.37}$$

Problems

1. Show that α_2 is also a solution of Eq. (3.1).
2. Under what conditions are the two solutions of (3.1) the same?
3. What are the values of A and B in

$$(A - 4)e^{\alpha_1 t} + Be^{\alpha_2 t} = 0, \qquad \alpha_1 \neq \alpha_2$$

4. Using similar logic, find A and B if

$$(A-5)t^2 = (A+B)e^{-5t}$$

5. Verify that both of the following are solutions of

$$\frac{d^2C}{dt^2} = K_1$$

(a) $C = \dfrac{K_1 t^2}{2} + 4t + 5$ (b) $C = \dfrac{K_1 t^2}{2} + 5t + 3$

How do these solutions differ in their behavior at $t = 0$ and how do their derivatives differ?

6. Show that

$$C = \frac{K_1 t^2}{2} + A_1 t + A_2$$

is not a solution of

$$\frac{d^2C}{dt^2} = K_1 C$$

unless $K_1 = 0$.

7. Given that C_1 is a solution of

$$\frac{d^2C_1}{dt^2} + K_1\frac{dC_1}{dt} + K_2 C_1 = 0$$

show that $C_2 = C_1 + (K_3/K_2)$ is a solution of

$$\frac{d^2C_2}{dt^2} + K_1\frac{dC_2}{dt} + K_2 C_2 = K_3$$

Do not do this by using the solutions given in Table III-1.

8. By differentiating, show that each of the following is by itself a solution of

$$\frac{d^2C}{dt^2} + K_1\frac{dC}{dt} + K_2 C = 0$$

if $K_1{}^2 = 4K_2$:

(a) $C_1 = A_1 e^{-(K_1/2)t}$ (b) $C_2 = A_2 t e^{-(K_1/2)t}$

If each of these is individually a solution, how can you show that

$$C_3 = C_1 + C_2$$

is also a solution?

9. Which of the following are linear? Which are homogeneous? What is the order of each equation?

(a) $\dfrac{dC(t)}{dt} = C(t)$ (b) $\dfrac{dC(t)}{dt} = C^2(t)$

(c) $\dfrac{d^2C(t)}{dt^2} + t\dfrac{dC(t)}{dt} + 1 = 0$ (d) $\dfrac{d^3C(t)}{dt^3} + C(t)\dfrac{dC(t)}{dt} = 0$

10. Which of the following will have exponential solutions?

(a) $\dfrac{d^2C(t)}{dt^2} + 4\dfrac{dC(t)}{dt} = 0$ (b) $\dfrac{d^2C(t)}{dt^2} + t\dfrac{dC(t)}{dt} = 0$

(c) $\dfrac{d^2C(t)}{dt^2} + 4\dfrac{dC(t)}{dt} + C(t) = 0$ (d) $\dfrac{d^4C(t)}{dt^4} + 4\dfrac{d^3C(t)}{dt^3} + \dfrac{dC(t)}{dt} = 0$

11. Solve the following equations.

(a) $\dfrac{dC(t)}{dt} = 4C(t), \qquad C(0) = 2$

(b) $\dfrac{dC(t)}{dt} = 4C(t) + 1, \qquad C(0) = 0$

(c) $\dfrac{d^2C(t)}{dt^2} = 0, \qquad C(1) = 2, \quad C(2) = 3$

(d) $\dfrac{d^2C(t)}{dt^2} = 4, \qquad C(1) = 5, \quad C(2) = 12$

12. Verify that Eq. (3.11) is a solution of (3.10).

13. In going from Eq. (3.14) to (3.15), Eq. (3.6) was used. What assumption was implicit in this? Had V_1 been equal to V_2, could this assumption have been made?

14. Show that

$$C_2 = Ate^{-\alpha t}$$

is a solution of (3.12) if V_1 equals V_2. Does it satisfy the accessory conditions of the problem? Finish the solution.

15. Verify the choice of β in the solution of (3.16).

16. Verify (3.36) and (3.37).

17. Suppose one places a beaker of water, initially at 20°C, on a hot plate at temperature 80°C. Sketch the temperature of the water as a function of time, given that it reaches 50°C in 10 min. When does it reach 65°C? Note: You do not really have all the data you might need to solve this problem, but make a reasonable guess based on the fact that the rate of heat loss from the beaker is proportional to the difference in temperature between the beaker and the room.

18. A patient was given 5 microcuries (μCi) of ^{131}I. Two hours later 0.5 μCi had been taken up by his thyroid. How much would have been taken up by the thyroid in 2 hours if he had been given 15 μCi? The important thing to know here is that the uptake and loss of ^{131}I is a linear process.

19. Is the blood sugar content linearly related to sugar intake? Can an argument similar to that involved in Problem 18 be made?

IV

The Laplace Transform

1. Introduction: Antiderivatives

In elementary calculus one is often taught that the process of integration is the reverse of the process of differentiation. However, the term "integral" is often used to refer to the area under a curve. This results in confusion between two quite different things, one of which is the indefinite integral and the other the definite integral, which is in fact the area under a curve. In some of the more recent mathematics texts this difficulty is avoided by use of the terms "antiderivative" and "antidifferentiation," the latter referring to the process that is the reverse of taking a derivative. We shall adhere to this nomenclature. When the term integral is used, it will always refer to the definite integral, which is explained below.

Given a function $f(x)$, its antiderivative, which we call $H(x)$, is that function which, when differentiated, yields $f(x)$. Thus, one antiderivative of x^2 is $\frac{1}{3}x^3$; however, another antiderivative of x^2 is $\frac{1}{3}x^3$ plus any constant, since the derivative of a constant is zero. This will always be the case in antidifferentiation. A constant can always be added to the antiderivative. Following are some examples of functions and their antiderivatives. Cover the right-hand side of the page and see if you can reconstruct the table of antiderivatives.

$\int 2\,dx$	$2x$ + constant
$\int x\,dx$	$\frac{1}{2}x^2$ + constant

$\int x^n\,dx$	$\frac{1}{n+1}x^{n+1}$ + constant
$\int \alpha e^{\alpha x}\,dx$	$e^{\alpha x}$ + constant
$\int \frac{1}{2} + \frac{3}{2}x + x^3\,dx$	$\frac{1}{2}x + \frac{3}{4}x^2 + \frac{1}{4}x^4$ + constant

The symbol for antidifferentiation is the elongated S sign $\int$. The symbol dx indicates the variable being integrated. Thus

$$\int xy\,dx = \tfrac{1}{2}x^2y + C$$

but

$$\int xy\,dy = \tfrac{1}{2}xy^2 + C$$

Note that the variable that is not being integrated is treated as if it were a constant.

2. Area under a Curve

Let us see how antidifferentiation is related to the area under a curve. Let

$$y = f(x)$$

be a function that is greater than zero and continuous between $x = x_0$ and $x = x_n$. Now let us define a new function $A(x)$, which is the area above the x axis and below the curve y, and to the right of x_0 up to x_v, where the subscript v indicates that this x will be treated as a variable. Thus, $A(x_v)$ is the crosshatched area in Figure 4-1. If $x_v = x_0$, then $A(x_v) = 0$. If $x_v = x_n$, then $A(x_v)$ is the entire area between x_0 and x_n.

Now let us divide the range x_0 to x_n into narrow strips, each of width Δx (Figure 4-2), and let us calculate the area between x_2 and x_3:

$$\text{area between } x_2 \text{ and } x_3 \approx f(x_2)\,\Delta x$$

where the symbol $\approx$ means approximately equal.

Actually, $f(x_2)\,\Delta x$ is the shaded rectangle, which differs from the exact area by the small triangular area above the rectangle, but this can be made arbitrarily small by choosing Δx small enough. Now the area from x_0 to x_2 is $A(x_2)$. Similarly, the area from x_0 to x_3 is $A(x_3)$, which is the same as $A(x_2 + \Delta x)$. The difference between these is the shaded rectangle. Therefore,

$$A(x_2 + \Delta x) - A(x_2) = \Delta x f(x_2)$$

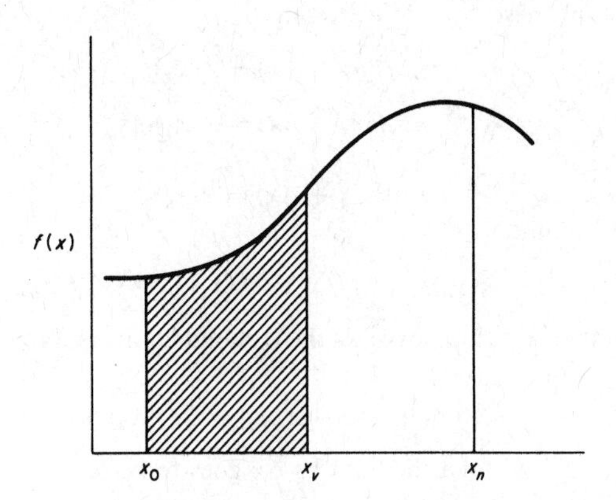

Figure 4-1. The crosshatched area represents the area under a curve defined as that area between the curve and the horizontal axis and between the given values of the independent variables x_0 and x_v.

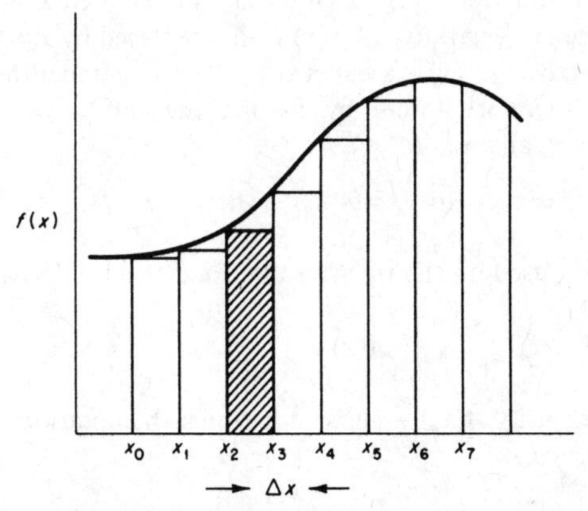

Figure 4-2. How the area under a curve can be approximated by a series of rectangular areas.

We can do the same thing for all the rectangles

$$A(x_0 + \Delta x) - A(x_0) = f(x_0)\,\Delta x$$

$$A(x_1 + \Delta x) - A(x_1) = f(x_1)\,\Delta x$$

$$A(x_2 + \Delta x) - A(x_2) = f(x_2)\,\Delta x$$

$$\vdots$$

$$A(x_{n-1} + \Delta x) - A(x_{n-1}) = f(x_{n-1})\,\Delta x$$

Now add up these equations, noting that

$$A(x_0 + \Delta x) = A(x_1)$$
$$A(x_1 + \Delta x) = A(x_2)$$
$$\vdots$$
$$A(x_{n-1} + \Delta x) = A(x_n)$$

The sum becomes

$$A(x_n) - A(x_0) = \Delta x[f(x_0) + f(x_1) + \cdots + f(x_{n-1})]$$

where the right side is the sum of all the rectangular areas. Now let us evaluate $A(x_0)$ from

$$A(x_0 + \Delta x) - A(x_0) = \Delta x f(x_0)$$

Divide by Δx and take the limit as Δx goes to zero.

$$\frac{dA(x_0)}{dx} = \lim_{\Delta x \to 0} \frac{A(x_0 + \Delta x) - A(x)}{\Delta x} = f(x_0)$$

$A(x_0)$ is that function $A(x)$ which when differentiated and evaluated at x_0 is $f(x_0)$. It is the antiderivative of $f(x)$ with x replaced by x_0. Similarly, $A(x_n)$ is the antiderivative of $f(x)$ evaluated at x_n. The area under the curve between x_0 and x_n is therefore found by finding the antiderivative of $f(x)$ and evaluating it at x_n and x_0.

$$\text{area } x_0 \text{ to } x_n = A(x_n) - A(x_0) = \int_{x_0}^{x_n} f(x)\, dx$$

The notation used on the right means: Find the antiderivative of $f(x)$

$$A(x) = \int f(x)\, dx$$

Evaluate the quantity $A(x_n) - A(x_0)$, for which the notation

$$A(x)\Big|_{x_0}^{x_n}$$

is usually used. We must be careful about one more point. We might have added a constant to each antiderivative. Thus, we might have gotten

$$\text{area } x_0 \text{ to } x_n = (A(x_n) + C_n) - (A(x_0) + C_0)$$

However, we can assure ourselves that $C_n - C_0$ is zero by moving x_n left until it falls on top of x_0. When we do so, the area must be zero.

Therefore, if

$$x_n = x_0, \qquad A(x_n) + C_n - A(x_0) - C_0 = 0$$

then $C_n = C_0$. The proof is now complete.

Let us summarize:

$$\text{area } x_0 \text{ to } x_n = \int_{x_0}^{x_n} f(x)\,dx = A(x)\Big|_{x_0}^{x_n} = A(x_n) - A(x_0)$$

where

$$A(x) = \int f(x)\,dx$$

Note that $A(x)$ is the antiderivative of $f(x)$, which is also called the indefinite integral of $f(x)$, but the area is the definite integral

$$\int_{x_0}^{x_n} f(x)\,dx = A(x)\Big|_{x_0}^{x_n}$$

The units of an integral are those of the function being antidifferentiated multiplied by the units of the variable being integrated. Thus, the height of a curve which has linear dimensions when integrated yields an area.

3. The Split Integral Theorem

Occasionally we wish to evaluate an integral in separate parts. We may do so by dividing the range of integration at any arbitrary point and noting that the sum of the areas to the left and right of the arbitrary division is the total area.

Thus

$$\int_{x_0}^{x_3} f(x)\,dx + \int_{x_3}^{x_n} f(x)\,dx = \int_{x_0}^{x_n} f(x)\,dx$$

In the proof above, the function $f(x)$ was assumed to be positive between x_0 and x_n. If, however, $f(x)$ becomes negative in any part of this range, the effect is to subtract from the definite integral the area that lies below the x axis but above the negative part of $f(x)$. We can see this by the following. In Figure 4-3, $f(x)$ is negative between x_4 and x_7:

$$A_{0,4} = \int_{x_0}^{x_4} f(x)\,dx, \qquad A_{4,7} = \int_{x_4}^{x_7} f(x)\,dx, \qquad A_{7,n} = \int_{x_7}^{x_n} f(x)\,dx$$

According to the split integral theorem, the sum of these three must be the integral from x_0 to x_n,

$$\int_{x_0}^{x_n} f(x)\,dx = A_{0,4} + A_{4,7} + A_{7,n}$$

However, since $f(x)$ is negative between x_4 and x_7, the area calculated for it will also be negative. Thus we find the result that the integral of a function

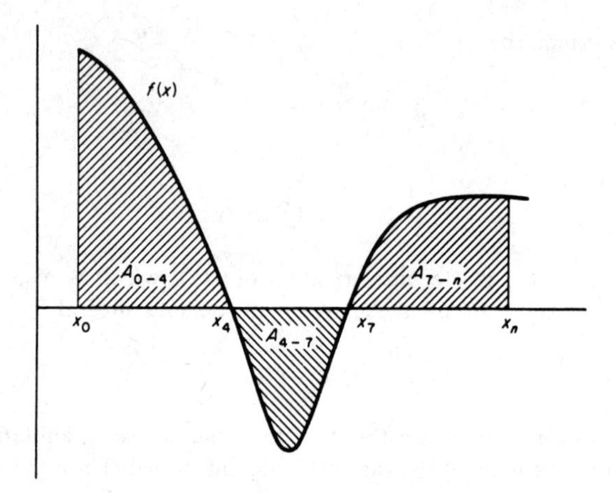

Figure 4-3. Definition of area under a curve when part of the curve goes below the horizontal axis.

that becomes negative in part of its range is the sum of the areas over the positive range of the function minus the sum of the areas over the negative range.

In elementary calculus considerable time is devoted to learning how to find antiderivatives of complicated functions. By the time most students are in graduate school they have completely forgotten how to do this, and in fact it is rarely necessary. Most functions that one wishes to antidifferentiate can be found in tables and the reader should perhaps practice finding the antiderivatives of the following functions and finding areas underneath curves as indicated:

$$\int_1^4 3x^2\,dx = 63$$

$$\int_1^5 4\,dx = 16$$

$$\int_1^2 \frac{x\,dx}{\sqrt{2x^2+3}} = \frac{1}{2}\left(\sqrt{11}-\sqrt{5}\right)$$

$$\int_1^2 e^{-\alpha x}\,dx = \frac{1}{\alpha}(e^{-\alpha} - e^{-2\alpha})$$

We must introduce one more concept, integrals with an infinite upper limit. It may seem paradoxical that one can define an area that goes from a fixed value of x to a point infinitely far to the right! In general, it does cause difficulty. However, if the function $f(x)$ gets small fast enough as x increases, integrals of this type can have finite value. We refer to these cases as convergent integrals, and for the purposes of this book, if the antiderivative evaluated for an infinitely large value of x has a finite value, the integral may be assumed to be convergent. Thus the following integrals are convergent:

$$\int_1^\infty \frac{1}{x^2}\,dx = -\frac{1}{x}\bigg|_1^\infty = 1$$

$$\int_0^\infty e^{-\alpha t}\,dt = -\frac{1}{\alpha}e^{-\alpha t}\bigg|_0^\infty = -\frac{1}{\alpha}(0-1) = \frac{1}{\alpha} \qquad \text{if} \quad \alpha > 0$$

But these integrals are not convergent:

$$\int_1^\infty \frac{1}{x}\,dx = \log x\bigg|_1^\infty = \log \infty - \log 1 = \infty$$

$$\int_0^\infty x\,dx = \tfrac{1}{2}x^2\bigg|_0^\infty = \infty - 0$$

Do Problems 1–3 at the end of this chapter.

4. Simultaneous Linear Equations

In connection with the partial fraction expansions that we shall use in this chapter we shall need to know how to solve simultaneous linear equations. This is a topic of high school algebra but the methods usually taught are not the best ways to handle numerical problems. The method illustrated here, called the Gauss reduction, involves much less labor. Here is a set of simultaneous linear equations:

$$\begin{aligned} 2A + 2B + C &= 14 \\ A - B + C &= 3 \\ 2A + 4B - 2C &= 8 \end{aligned} \tag{4.1}$$

It is simultaneous because all the equations must be solved together to get values for A, B, and C. It is a linear system because the unknowns A, B, C appear only in the first powers. An illustration of the technique will be self-explanatory. We begin by dividing each equation by the coefficient of A of that equation

$$\begin{aligned} A + B + \tfrac{1}{2}C &= 7 \\ A - B + \phantom{\tfrac{1}{2}}C &= 3 \\ A + 2B - \phantom{\tfrac{1}{2}}C &= 4 \end{aligned}$$

Now subtract the bottom equation from the two equations above it. This yields two equations from which A has been eliminated,

$$\begin{aligned} -B + \tfrac{3}{2}C &= 3 \\ -3B + 2C &= -1 \end{aligned} \tag{4.2}$$

Now divide each of these by the coefficient of B:

$$B - \tfrac{3}{2}C = -3$$

$$B - \tfrac{2}{3}C = \tfrac{1}{3}$$

Subtract the second from the first to find an equation from which both B and A have been eliminated:

$$-\tfrac{5}{6}C = -\tfrac{10}{3}$$

Thus

$$C = 4$$

Now, go back to (4.2) and insert the value of C, which is now known, and solve for B:

$$-B + \tfrac{3}{2}4 = 3$$

$$B = 3$$

Then put the values of B and C into (4.1) and find A:

$$2A + 2 \cdot 3 + 4 = 14$$

$$A = 2$$

Check the result by testing it

$$\begin{aligned} 2 \cdot 2 + 2 \cdot 3 \quad + 4 &= 14 \\ 2 \quad - 3 \quad + 4 &= 3 \\ 2 \cdot 2 + 4 \cdot 3 - 2 \cdot 4 &= 8 \end{aligned}$$

One often requires the solution of simultaneous equations that are symbolic rather than numerical. In the following, M_1, M_2, M_3, M_4, and N_1, N_2 are given and the unknowns are B_1 and B_2:

$$M_1 B_1 + M_2 B_2 = N_1$$

$$M_3 B_1 + M_4 B_2 = N_2$$

Divide the first equation by M_1 and the second by M_3

$$B_1 + \frac{M_2}{M_1} B_2 = \frac{N_1}{M_1}$$

$$B_1 + \frac{M_4}{M_3} B_2 = \frac{N_2}{M_3}$$

$$B_2\left(\frac{M_2}{M_1} - \frac{M_4}{M_3}\right) = \left(\frac{N_1}{M_1} - \frac{N_2}{M_3}\right)$$

To find B_1, divide the first equation of the simultaneous pair by M_2 and the second by M_4

$$\frac{M_1}{M_2} B_1 + B_2 = \frac{N_1}{M_2}$$

$$\frac{M_3}{M_4} B_1 + B_2 = \frac{N_2}{M_4}$$

$$\left(\frac{M_1}{M_2} - \frac{M_3}{M_4}\right) B_1 = \left(\frac{N_1}{M_2} - \frac{N_2}{M_4}\right)$$

Thus

$$B_1 = \left(\frac{N_1}{M_2} - \frac{N_2}{M_4}\right) \Big/ \left(\frac{M_1}{M_2} - \frac{M_3}{M_4}\right) = \frac{N_1 M_4 - N_2 M_2}{M_1 M_4 - M_2 M_3}$$

$$B_2 = \left(\frac{N_1}{M_1} - \frac{N_2}{M_3}\right) \Big/ \left(\frac{M_2}{M_1} - \frac{M_4}{M_3}\right) = \frac{N_2 M_1 - N_1 M_3}{M_1 M_4 - M_2 M_3}$$

Do Problems 4 and 5 at the end of this chapter.

5. Transforming the Differential Equation

Although the reader is strongly encouraged to develop the knack of guessing the solutions of simple linear differential equations, there will inevitably arise cases in which he cannot guess correctly. In many of these cases, the value of the dependent variable and its derivatives will be known at time zero or at x equal to zero. For problems of this type, the Laplace transform is almost invariably the technique of choice. It allows very complicated differential equations to be reduced to almost mechanical algebraic manipulation.

The Laplace transform, at the level at which we will use it in this book, consists of a set of rules and tables that allow one to perform a pure "turn the crank" solution of a differential equation and its accessory conditions. It may appear a bit mysterious, and indeed did so to mathematicians for many years until its underlying rigorous basis was found. In spite of the mysterious aspect, the Laplace transform is so powerful a tool, as compared to other methods of solving differential equations, that its inclusion in a book of this type is absolutely necessary. The rules of the game are as follows.

Appendix II is a table of functions $G(t)$ where t is time or any other independent variable. Associated with each $G(t)$ is a function $g(s)$ which is called the *Laplace transform* of $G(t)$.† The role of the new variable s will appear shortly. The following notation is often used to indicate that $g(s)$ is the Laplace transform of $G(t)$.

$$g(s) = L[G(t)]$$

This notation is used to indicate that $G(t)$ is the function whose transform is $g(s)$:

$$G(t) = L^{-1}[g(s)]$$

To solve a linear differential equation in which the accessory conditions are known at time zero (or the zero value of whatever is the independent variable), proceed as follows.

1. When $G(t)$ appears in the equation undifferentiated, replace it by $g(s)$. When the first derivative of G appears (dG/dt), replace it by $sg(s) - G(0)$, where $G(0)$ is to be evaluated at time zero, or at some infinitesimally greater time, if G changes at time zero. For example, if G represents a concentration and a quick injection is made at time zero, G is evaluated after the injection. When the second derivative appears (d^2G/dt^2), replace it by

$$s^2g(s) - sG(0) - \frac{dG(0)}{dt}$$

where both $G(0)$ and $dG(0)/dt$ are evaluated after any change that occurs at time zero. If higher-order derivatives appear (d^nG/dt^n), replace them by

$$s^ng(s) - s^{n-1}G(0) - s^{n-2}\frac{dG(0)}{dt} - s^{n-3}\frac{d^2G(0)}{dt^2} - \cdots - \frac{d^{n-1}G(0)}{dt^{n-1}}$$

If any time-dependent functions appear (functions of the independent variable), replace them by their Laplace transform from the table, or by means of formula given later in this chapter. This process is called transforming the differential equation.

2. The resulting transformed equation contains no derivatives. It is a

† Many books use the letters F or $F(t)$ and f or $f(s)$ for the Laplace transform.

purely algebraic equation that can be solved for $g(s)$. Find $g(s)$. Look up the corresponding $G(t)$ in the table (Appendix II). This is the solution of the differential equation. Check it by differentiating and inserting it in the original equation.

The power of the Laplace transform lies in the fact that it removes derivatives from differential equations and replaces them with algebraic quantities in the transformed equations. If the transformed equation is to be valid, however, it must be dimensionally correct if the equation that was transformed was dimensionally correct. We are going to replace first derivatives of G with respect to time by $sg - G(0+)$. Therefore sg must have the same units as G. We are going to replace second derivatives with respect to time by

$$s^2g - sG(0+) - \frac{dG(0)}{dt}$$

Thus s must have units of reciprocal time. Therefore, g has units of time multiplied by the units of $G(t)$.

If we had been replacing derivatives with respect to a length dimension, g would have units of length times the units of G and s would have units of reciprocal length (length^{-1}).

6. One Compartment with Nonzero Initial Concentration

A few examples will make this clear. Let us return to the one-compartment dilution with the constant input R of tracer, but with the added complication that the initial concentration may not be zero. The differential equation (2.9) was

$$V\frac{dC(t)}{dt} + FC(t) = R$$

and the accessory condition is the value of $C(0)$. We note that the differential equation is of the linear inhomogeneous type. In this case the concentration $C(t)$ takes the role of $G(t)$.

From Appendix II, entry 1, we find that the transform of an additive constant 1 is $1/s$. The transform of R (a constant) is therefore R/s. Replace the time derivative of C by $sg(s) - C(0)$. Thus the transformed equation is

$$V(sg(s) - C(0)) + Fg(s) = \frac{R}{s}$$

Solve for g and express it as a sum of the standard forms given in Appendix II.

$$(Vs + F)g(s) = VC(0) + \frac{R}{s}$$

$$g(s) = \frac{C(0)}{s + (F/V)} + \frac{\frac{R}{V}}{s(s + F/V)}$$

These two forms are found in Appendix II as entries 5 and 6, where $a = F/V$. Thus

$$C(t) = C(0)e^{-(F/V)t} + \frac{R}{V}\frac{V}{F}(1 - e^{-(F/V)t})$$

This is a more general solution than the one we found previously. In this case the initial value of $C(0)$ is arbitrary, whereas in the previous case we assumed that it was zero. If we set it to zero, this solution is naturally found to agree with that worked out in the last chapter. Note the simplicity of the method. In a wide variety of cases it reduces the solution of linear differential equations to a purely mechanical operation.

7. The Second Compartment of the Two-Compartment Series Dilution

Let us try a second example, the second compartment of the two-compartment series dilution problem. The differential equation that we derived previously as (2.13) was

$$\frac{dC_2(t)}{dt} + \frac{F}{V_2}C_2(t) = \frac{FC_1(0)}{V_2}e^{-(F/V_1)t}$$

with the accessory condition that $C_2(0) = 0$.

We proceed as in the previous case, replacing $C(t)$ by $g(s)$ and replacing its derivative by $sg(s) - C_2(0)$. On the right-hand side of the equation we have a function of time. We search the transform table (Appendix II) until we find a form that is similar to the time-dependent term, in this case entry 5, with $a = F/V_1$, and replace it by its transform. The resulting transformed equation is

$$sg(s) - C_2(0) + \frac{F}{V_2}g(s) = \frac{FC_1(0)}{V_2}\frac{1}{s + (F/V_1)}$$

Let us for simplicity immediately insert the initial condition on $C_2(0)$, which is that it be equal to zero shortly after time zero. Thus our equation simplifies to the form

$$g(s) = \frac{C_1(0)F}{V_2} \frac{1}{s + (F/V_1)} \frac{1}{s + (F/V_2)}$$

which we easily solve for g and invert (the process of going from the Laplace transform equation back to the time-dependent equation) by means of entry 11 in Appendix II,

$$C_2(t) = \frac{V_1 C_1(0)}{V_2 - V_1} [e^{-(F/V_2)t} - e^{-(F/V_1)t}]$$

8. Diffusion between Compartments

Let us consider the two-compartment diffusion problem of Chapter III and solve it by two different techniques as examples of the use of the Laplace transform. The differential equations (3.22) to be solved are

$$V_1 \frac{dC_1(t)}{dt} = K(C_2 - C_1),$$
$$V_2 \frac{dC_2(t)}{dt} = K(C_1 - C_2) \tag{4.3}$$

where $C_1(0)$ and $C_2(0)$ are known.

The first method depends upon eliminating one variable, in this case C_2, from the pair of equations. This is done as follows. The first equation is solved for C_2

$$C_2(t) = C_1(t) + \frac{V_1}{K} \frac{dC_1(t)}{dt}$$

It is then differentiated once, noting that the derivative of a derivative is a second derivative, as explained in the introduction to Chapter III:

$$\frac{dC_2(t)}{dt} = \frac{dC_1(t)}{dt} + \frac{V_1}{K} \frac{d^2C_1(t)}{dt^2}$$

The value of C_2 and its derivative are then inserted in the second equation of the pair of differential equations (4.3)

$$V_2\left[\frac{V_1}{K}\frac{d^2C_1}{dt^2} + \frac{dC_1}{dt}\right] = K\left[C_1 - \left(C_1 + \frac{V_1}{K}\frac{dC_1}{dt}\right)\right]$$

and collecting terms we have

$$\frac{d^2C_1(t)}{dt^2} + K\left(\frac{1}{V_1} + \frac{1}{V_2}\right)\frac{dC_1}{dt} = 0$$

which is a second-order differential equation but with only one unknown C_1. This is now solved by means of the Laplace transform. Let us transform the equation and solve for g:

$$s^2g(s) - sC_1(0) - \frac{dC_1(0)}{dt} + K\left(\frac{1}{V_1} + \frac{1}{V_2}\right)(sg(s) - C_1(0)) = 0$$

$$g = \frac{C_1(0)}{s+a} + \frac{aC_1(0) + [dC_1(0)/dt]}{s(s+a)}$$

where

$$a = K\left(\frac{1}{V_1} + \frac{1}{V_2}\right)$$

The transform is then inverted by means of entries 5 and 6 in Appendix II:

$$C_1(t) = C_1(0)e^{-at} + \left(aC_1(0) + \frac{dC_1(0)}{dt}\right)\frac{1}{a}(1 - e^{-at})$$

In order to determine the value of the derivative at time 0, we go back to the original equations (4.3), where we find from the first one that

$$V_1\frac{dC_1(0)}{dt} = K(C_2(0) - C_1(0))$$

We can insert this value into the previous equation and with a little additional manipulation get it into the same form as is given in (3.32):

$$C_1(t) = C_1(0)e^{-at} + C_1(0)(1 - e^{-at}) + \frac{1}{a}(1 - e^{-at})\frac{dC_1(0)}{dt}$$

$$= C_1(0) + \frac{V_2}{V_1 + V_2}(C_2(0) - C_1(0))(1 - e^{-at})$$

We might have solved the same problem in a different way. Instead of eliminating one concentration, we can instead perform two different Laplace transforms, g_1 the transform of C_1 and g_2 the transform of C_2. We may then eliminate one transform between the resulting pair of equations as follows. Transforming the pair (4.3), we find

$$V_1(sg_1 - C_1(0)) = K(g_2 - g_1)$$

$$V_2(sg_2 - C_2(0)) = K(g_1 - g_2)$$

From the second of these we get

$$g_2 = \frac{Kg_1 + V_2 C_2(0)}{sV_2 + K}$$

which we insert into the first:

$$V_1 s g_1 - V_1 C_1(0) = \frac{K^2 g_1 + KV_2 C_2(0)}{sV_2 + K} - Kg_1$$

$$g_1 = \frac{sV_1 V_2 C_1(0) + V_1 K C_1(0) + V_2 K C_2(0)}{s^2 V_1 V_2 + KV_1 s + KV_2 s}$$

$$= \frac{C_1(0)}{s+a} + K \frac{(C_1(0)/V_2) + (C_2(0)/V_1)}{s(s+a)}$$

where

$$a = K \frac{V_1 + V_2}{V_1 V_2}$$

Inverting these, we find

$$C_1(t) = C_1(0)e^{-at} + \frac{K}{a}\left(\frac{C_1(0)}{V_2} + \frac{C_2(0)}{V_1}\right)(1 - e^{-at})$$

Now

$$\frac{K}{a}\left(\frac{C_1(0)}{V_2} + \frac{C_2(0)}{V_1}\right) = \frac{V_1 C_1(0) + V_2 C_2(0)}{V_1 + V_2}$$

$$= C_1(0) + (C_2(0) - C_1(0)) \frac{V_2}{V_1 + V_2}$$

Therefore

$$C_1(t) = C_1(0) + (1 - e^{-at})(C_2(0) - C_1(0)) \frac{V_2}{V_1 + V_2}$$

The latter method of eliminating transformed variables rather than the original variables is very commonly used in systems involving more than two unknowns because of the difficulty of removing derivatives in multiple unknown situations.

9. Finding New Transforms

It must be clear that the Laplace transform is an extremely powerful and useful tool. It is of course limited to a relatively narrow class of differential equations, but a class that occurs with great frequency in biological and incidentally, physical problems. Unfortunately, one often needs transforms and inverse transforms that are not in the available tables. Transforms are usually fairly easy to calculate from the following integral, which is actually the definition of the Laplace transform,

$$g(s) = \int_0^\infty e^{-st} G(t)\, dt$$

Taking, for example, one of the entries in Appendix II, we find the transform of e^{-at} from

$$g(s) = \int_0^\infty e^{-st} e^{-at}\, dt = \int_0^\infty e^{-(s+a)t}\, dt = \frac{1}{s+a}$$

Example. This example is not in the tables. Find the transform of $\sqrt{t}$.

$$g(s) = \int_0^\infty e^{-st} \sqrt{t}\, dt$$

Let $r = \sqrt{t}$ or $t = r^2$; $dt = dr\, 2r$

$$g(s) = \int_0^\infty e^{-sr^2} r 2r\, dr = 2 \int_0^\infty r^2 e^{-sr^2}\, dr$$

This form is found in most standard integral tables as

$$\int_0^\infty x^{2n} e^{-ax^2}\, dx = \frac{1 \cdot 3 \cdot 5 \cdot 7 \cdots (2n-1)}{2^{n+1} a^n} \sqrt{\frac{\pi}{a}}$$

Therefore, by choosing n equal to one and a equal to s, we get

$$g(s) = \frac{1}{2} \frac{\sqrt{\pi}}{s^{3/2}}$$

Thus the forward transforms can usually be found with a set of integral tables and some backbreaking labor. Getting the inverse of a transform is frequently more difficult. There exists a very powerful technique, called *contour integration*, which can be used for a wide variety of transform inversions, but it requires a fairly sophisticated knowledge of complex variables and is beyond the range of this book. There are, however, other methods that can be used fairly easily. We shall describe two such methods later in this chapter, but before doing so, we take note of three special cases that occur with particular frequency in the remainder of this book.

10. Three Special Cases

To find the inverses of transforms of the type

$$\frac{1}{s^2 + ps + q}, \qquad \frac{s}{s^2 + ps + q}, \qquad \frac{1}{s(s^2 + ps + q)} \tag{4.4}$$

in which p^2 is greater than $4q$ and q is not zero (q equal to zero reduces to entries 5, 6, and 7 in Appendix II), we note that the parentheses in the denominators can be factored into

$$s^2 + ps + q = (s + a)(s + b)$$

if

$$a + b = p, \qquad ab = q$$

We can solve this pair of equations for a and b by letting

$$a = \frac{q}{b}, \qquad b = \frac{q}{a}$$

from which we get

$$\frac{q}{b} + b = p, \qquad a + \frac{q}{a} = p$$

$$b^2 - pb + q = 0, \qquad a^2 - pa + q = 0$$

Thus, a and b are the roots of a quadratic equation

$$x^2 - px + q = 0$$

We choose a to have the positive sign before the square root.†

$$a = \tfrac{1}{2}(p + \sqrt{p^2 - 4q}), \qquad b = \tfrac{1}{2}(p - \sqrt{p^2 - 4q}), \qquad ab = q, \qquad a + b = p$$

In terms of a and b the inverse transforms are

$$L^{-1}\left[\frac{1}{s^2 + ps + q}\right] = \frac{e^{-at} - e^{-bt}}{b - a}$$

$$L^{-1}\left[\frac{s}{s^2 + ps + q}\right] = \frac{be^{-bt} - ae^{-at}}{b - a}$$

$$L^{-1}\left[\frac{1}{s(s^2 + ps + q)}\right] = \frac{1}{ab}\left[1 + \frac{ae^{-bt} - be^{-at}}{b - a}\right]$$

$$= \frac{1}{a - b}\left[\frac{1}{b}(1 - e^{-bt}) - \frac{1}{a}(1 - e^{-at})\right]$$

† If $p^2 = 4q$, then $a = b$ and the fractions in the above become zero in both their numerators and denominators. There exist ways of handling this, but in most cases it is better to factor the denominator of the transform as follows and use entries 8, 9, and 10 of Appendix II. If $p^2 = 4q$, then $s^2 + ps + q = [s + (p/2)]^2$.

Let us verify the third of the foregoing inverse transforms.

$$L\left[\frac{1}{ab}\left(1+\frac{ae^{-bt}-be^{-at}}{b-a}\right)\right]=\frac{1}{ab}\left[\frac{1}{s}+\frac{1}{b-a}\left(\frac{a}{s+b}-\frac{b}{s+a}\right)\right]$$

$$=\frac{1}{ab}\left(\frac{1}{s}+\frac{1}{b-a}\left(\frac{as+a^2-bs-b^2}{s^2+(a+b)s+ab}\right)\right)$$

$$=\frac{1}{q}\left(\frac{1}{s}-\frac{s+(a+b)}{s^2+(a+b)s+ab}\right)$$

$$=\frac{1}{q}\left(\frac{1}{s}-\frac{s+p}{s^2+ps+q}\right)=\frac{1}{s(s^2+ps+q)}$$

There will be a number of cases in which $4q/p^2$ is very small compared to 1. When this is true, we can make the approximation

$$\sqrt{p^2-4q}=p\sqrt{1-\frac{4q}{p^2}}=p\left(1-\frac{2q}{p^2}\right)=p-\frac{2q}{p}$$

which we justify by noting that

$$\left(1-\frac{2q}{p^2}\right)^2=1-\frac{4q}{p^2}+\left(\frac{2q}{p^2}\right)^2$$

but if $4q/p^2$ is small compared to 1, then $(2q/p^2)^2$ is very small compared to 1. Thus when $4q/p^2 \ll 1$,

$$a=p-\frac{q}{p}, \qquad b=\frac{q}{p} \tag{4.5}$$

The two other methods of evaluating inverse Laplace transforms that we shall discuss are the partial fraction expansion and the convolution integral. Although both of these methods are quite useful, it happens that they are not used often in this book, so that the reader can, if he wishes, skip these two sections for the moment, and return to them if the need arises.

11. The Partial Fraction Expansion

The partial fraction expansion is useful in those cases where the transform can be expressed as a quotient of polynomials and where these polynomials can be expressed as a sum of simpler transforms, as in

$$\frac{s^2+2}{s(s+2)(s+3)}=\frac{\frac{1}{3}}{s}-\frac{3}{s+2}+\frac{\frac{11}{3}}{s+3}$$

We can verify the expansion by calculating the sum

$$\frac{\frac{1}{3}}{s} - \frac{3}{s+2} + \frac{\frac{11}{3}}{s+3} = \frac{(s+2)(s+3)\frac{1}{3} - 3(s+3)s + \frac{11}{3}s(s+2)}{s(s+2)(s+3)}$$

$$= \frac{s^2(\frac{1}{3} - 3 + \frac{11}{3}) + s(\frac{5}{3} + \frac{22}{3} - 9) + \frac{6}{3}}{s(s+2)(s+3)}$$

$$= \frac{s^2+2}{s(s+2)(s+3)}$$

The inverse transform is

$$L^{-1}\left[\frac{\frac{1}{3}}{s}\right] - L^{-1}\left[\frac{3}{s+2}\right] + L^{-1}\left[\frac{\frac{11}{3}}{s+3}\right] = \frac{1}{3} - 3e^{-2t} + \frac{11}{3}e^{-3t}$$

The method of finding the partial fractions is to express the transform in terms of a sum of fractions in which the denominator of each partial fraction is a factor of the transform and the numerator is an unknown constant. The constants can then be determined by solving a set of simultaneous equations. The process is more easily exhibited than explained.

Let

$$\frac{s^2+2}{s(s+2)(s+3)} = \frac{A}{s} + \frac{B}{s+2} + \frac{C}{s+3}$$

$$= \frac{(A+B+C)s^2 + (5A+3B+2C)s + 6A}{s(s+2)(s+3)}$$

Therefore, by matching like powers of s, we obtain

$$A + B + C = 1$$

$$5A + 3B + 2C = 0$$

$$6A = 2$$

from which (see Section 4) we get

$$A = \tfrac{1}{3}, \qquad B = -3, \qquad C = \tfrac{11}{3}$$

12. The Convolution Integral

The other method of transform inversion that is commonly used involves the convolution integral. It is particularly useful when the transform can be represented as a product of two other transforms. Before introducing the convolution integral, it is necessary to digress for a moment to discuss the properties of parametric integrals, integrals that contain variable parameters other than the parameter being integrated.

Following are a number of examples of evaluated definite integrals. Notice that the variable that has been integrated never appears in the result of the evaluated integral. This is true for any definite integral, that is, an integral in which the limits are not functions of the variable being integrated:

$$\int_0^\infty \frac{\sin^2 x}{x^2}\, dx = \frac{\pi}{2}$$

$$\int_1^2 e^{3x}\, dx = \tfrac{1}{3}(e^6 - e^3)$$

$$\int_a^b t^2\, dt = \tfrac{1}{3}(b^3 - a^3)$$

A parametric definite integral is one in which some other variable appears either inside the integral or in one of the limits. Thus the following integrals are parametric. Note that in each case the parameter appears in the result but the variable being integrated does not appear.

$$\int_0^\infty \frac{e^{-ax}}{\sqrt{x}}\, dx = \sqrt{\frac{\pi}{a}}$$

$$\int_0^a t^2\, dt = \frac{a^3}{3}$$

$$\int_1^a e^{3x}\, dx = \tfrac{1}{3}\,(e^{3a} - e^3)$$

The reader should be sure to understand the foregoing examples of parametric integration, as the latter is the most difficult point in understanding the convolution integral.

The convolution theorem states that if a transform can be expressed as the product of two other transforms, its inverse can be expressed as either of two parametric integrals. If

$$g(s) = k(s)h(s)$$

then

$$G(t) = \int_0^t K(\tau)H(t-\tau)\, d\tau = \int_0^t K(t-\tau)H(\tau)\, d\tau \tag{4.6}$$

where K is the inverse transform of $k(s)$ and H is the inverse of $h(s)$. The notation $K(\tau)$ means replace t in the usual inverse by a new variable τ. and $H(t-\tau)$ means replace t by $t-\tau$.

Example. Suppose we did not have entry 11 in Appendix II. We could synthesize it by letting

$$g(s) = \frac{1}{(s+a)(s+b)}, \qquad k(s) = \frac{1}{s+a}, \qquad h(s) = \frac{1}{s+b}$$

The inverse of $k(s)$ is e^{-at} and we replace t by τ for the first form of the convolution. The inverse of $h(s)$ is e^{-bt} and we replace t by $t - \tau$:

$$G(t) = \int_0^t e^{-a\tau} e^{-b(t-\tau)}\, d\tau = \int_0^t e^{-(a-b)\tau} e^{-bt}\, d\tau$$

Note that e^{-bt} can be brought outside the integral since it does not contain the variable being integrated.

$$G(t) = e^{-bt} \int_0^t e^{-(a-b)\tau}\, d\tau = -e^{-bt} \frac{1}{a-b} (e^{-(a-b)t} - 1)$$

$$= \frac{1}{a-b} (e^{-bt} - e^{-at})$$

which is the same as entry 11, Appendix II.

Example. Let us try another example that is not in the table (Appendix II). Let us find the inverse of

$$g(s) = \frac{1}{(s+a)^3}$$

Looking in the table we find that we have the two transforms

$$\frac{1}{s+a}, \qquad \frac{1}{(s+a)^2}$$

Therefore, let

$$k(s) = \frac{1}{s+a}$$

the inverse of which is

$$e^{-at} \tag{4.7}$$

and

$$h(s) = \frac{1}{(s+a)^2}$$

the transform of which is

$$te^{-at} \tag{4.8}$$

Now in (4.7) replace t by $t - \tau$ and in (4.8) replace t by τ. Thus, according to the convolution theorem we have

$$G(t) = \int_0^t \tau e^{-a\tau} e^{-(t-\tau)a} \, d\tau = e^{-at} \frac{t^2}{2}$$

which the reader can verify. For further information on the Laplace transform see (6, 9).

Problems

1. If

$$f(x) = \tfrac{1}{2} x^2$$

what is the value of the integral

$$\int_1^3 f(x) \, dx$$

What are its dimensions if x is in centimeters?

2. Find

$$\int_0^\infty e^{-(F/V)t} \, dt$$

F is the flow, in liters per second; V, volume in liters; t, time in seconds. What are the units of the integral?

3. Sketch $y = (x - 1)(x - 3)$ from $x = 0$ to $x = 3$. Find

$$A = \int_0^1 y \, dx, \qquad B = \int_1^3 y \, dx, \qquad C = \int_0^3 y \, dx$$

Does C make sense?

4. Solve by Gauss reduction

$$2x + 4y + 6z = 28$$

$$3x - 3y + 3z = 6$$

$$3x + 3y + 6z = 27$$

5. Solve for x and y

$$Ax + By = C$$

$$Dx + Ey = F$$

What happens if C and F are both zero?

6. Solve by Laplace transform

$$\frac{dy}{dt} - ky = 0, \qquad y(0) = 3$$

7. Solve by Laplace transform

$$\frac{dy}{dt} - ky = \frac{t}{T^2}, \qquad y(0) = 2$$

Verify your solution by differentiating and checking that $y(0) = 2$.

8. Solve by Laplace transform

$$\frac{dC}{dt} + \frac{L}{V} C = Ae^{-at} + \frac{R}{V}, \qquad C(0) = 0$$

Verify the solution.

9. Solve by Laplace transform

$$\frac{d^2C(t)}{dt^2} - a^2C(t) = 0, \qquad C(0) = 1, \quad \frac{dC(0)}{dt} = 0$$

10. Repeat Problem 9 with the conditions

$$C(0) = 0, \qquad \frac{dC(0)}{dt} = 1$$

11. Solve the following pair of simultaneous differential equations by transforming and eliminating g_2.

$$\frac{dY_1}{dx} + KY_2 = R$$

$$\frac{dY_2}{dx} = KY_1$$

$$Y_1(0) = Y_2(0) = 0$$

Show

$$g_1 = \frac{R}{s^2 + K^2}$$

Find Y_1 and Y_2 and verify.

12. Solve

$$\frac{d^2x}{dp^2} + \frac{dx}{dp} + 0.09x = 0, \qquad x(0) = 1, \quad \frac{dx(0)}{dp} = 0$$

13. Making use of a table of integrals, verify the Laplace transforms 6 and 8 of Appendix II.

V

Compartmental Problems

1. Introduction

Compartmental problems are among the most frequently encountered applications of the methods discussed thus far in this book. They arise in biology in many different ways. In some problems, the compartments are physically divided, for example, digestive system to bloodstream to tissue, lungs to blood to tissue, extracellular space to intercellular space. In others they are not physically divided but rather may be chemically distinguishable pools that share the same space.

We have already discussed washout phenomena from one- and two-compartment series systems and simple diffusion between two connected compartments. In both of these cases the rate of transfer of the substance in question, which we usually refer to as a tracer, has been dependent upon its concentration in the compartment. While this is true in many cases, there are other cases in which the rate of removal of a substance from a compartment is not principally dependent upon its concentration in that compartment. For example, the rate at which oxygen is removed from blood within a given volume of tissue is largely independent of its partial pressure in the blood unless the partial pressure falls to extremely low levels. The reason for this, of course, is that the rate of oxygen consumption is a function of metabolic activity, which normally is not significantly influenced by the partial pressure.

Thus, in a discussion of even very simple compartmental problems, we must consider both concentration-dependent loss and concentration-independent loss. One must remember, of course, that both of these cases are extremes and that many reactions may be partially dependent on concentration.

We may also wish to consider a variety of inputs. We have already solved cases in which the input of the tracer occurred in an instantaneous bolus at time zero. We call this an *impulse input*. We have seen at least one case in which the tracer was *injected continuously, beginning at time zero*. We call this a *step input*, since when plotted against time, the rate of input appears as a step at time zero. We may wish to consider inputs that are functions of time, as was, for example, the input to the second compartment of a two-compartment dilution process.

Previously, we have considered loss of substances proportional to the concentration as being due to flow through the compartment. In the more general case, it may be due to flow through the compartment or it may be due to utilization within the compartment. Let us, therefore, use the letter L with the understanding that the quantity L multiplied by the concentration will be the rate of loss of the substance, which is proportional to concentration.

Let us also generalize the concept of input to incorporate concentration-independent loss. Let I be the rate of input of the substance to the compartment. If the compartment happens to have loss that is independent of concentration, this will be incorporated into I as a negative quantity. Strictly speaking, there are no reactions in which substances disappear exactly independently of concentration. There are, however, wide varieties of cases where, over physiological ranges, the disappearance of a substance is essentially independent of its concentration. We have already noted the case of oxygen. Another such case is that of active transport, where on one side of the membrane we see the substance disappearing at a rate essentially independent of concentration and appearing on the other side of the membrane at the same fixed rate.

2. The One-Compartment Mixed Case

We begin with a one-compartment process that we can think of as either a model of diffusion out of a cell embedded in an essentially constant environment, or as a chemical reaction whose velocity is determined by the concentration of the product and that of a reacting substance of fixed concentration.

In the case of a substance diffusing out of a cell, we find the differential equation that describes the process by considering the change in the amount of the diffusing substance within the cell that occurs during a small interval of time Δt. For generality, we assume that the substance is generated within the cell at a fixed rate I independent of concentration, remembering that if we wish to consider this concentration-independent loss within the cell, we do so by letting I be negative.

In a small interval of time Δt the change in the amount of the substance within the cell consists of the addition of an amount $I\,\Delta t$ and the loss of an amount $LC\,\Delta t$ where L is the diffusion constant or represents any other concentration-dependent loss. Therefore

$$\Delta Q = \Delta t(I - LC)$$

$$\text{moles} = \text{sec}\left(\frac{\text{moles}}{\text{sec}} - \frac{\text{moles/sec}}{\text{mole/cm}^3}\ \text{moles/cm}^3\right)$$

where, to ensure that the equation is dimensionally balanced, we have indicated the dimensions of each term beneath the equation. Now let us convert to a concentration change by noting that ΔC is equal to $\Delta Q/V$

$$\Delta C = \Delta t \frac{1}{V}(I - LC)$$

Dividing by Δt and taking the limit as Δt goes to zero, we find

$$\frac{dC}{dt} = \frac{1}{V}(I - LC) \tag{5.1}$$

If we wish to think of this as a chemical process in which substance A at fixed concentration A becomes substance C at concentration C, the rate of the forward reaction is $k_1 A$, while the rate of the reverse reaction is $k_2 C$ and the net change in product in a small interval of time is

$$\Delta Q = \Delta t\,(k_1 A - k_2 C)$$

or, in terms of the concentration of C,

$$\Delta C = \Delta t \frac{1}{V}(k_1 A - k_2 C),$$

$$\frac{dC}{dt} = \frac{1}{V}(k_1 A - k_2 C)$$

which is the same kind of equation as (5.1).

The reader should pause at this point and consider what he thinks the solution of Eq. (5.1) will be in three different cases: the steady-state case, in which I is a constant and the system comes to a steady-state concentration; the time-dependent case, in which C is zero at time zero but a steady-state input I begins at time zero; and the impulse case, where I is zero but C is not zero at time zero.

The steady-state case is trivial to analyze. In the steady state, dC/dt is zero, so that we find for the steady-state concentration

$$C_{ss} = \frac{I}{L}$$

The case of $C(0)$ zero with a constant rate of input starting at zero is easily solved by means of the Laplace transform. Noting that the transform of I/V is I/sV, we have

$$sg(s) - C(0) = \frac{I}{sV} - \frac{L}{V} g(s)$$

$$g(s) = \frac{I}{Vs\left(s + \frac{L}{V}\right)} \tag{5.2}$$

$$C(t) = \frac{I}{L}\left(1 - e^{-(L/V)t}\right)$$

The third case, with I equal to zero but $C(0)$ not zero, is also easily done by means of the Laplace transform:

$$sg(s) - C(0) = -\frac{L}{V} g(s)$$

$$g(s) = \frac{C(0)}{s + (L/V)} \tag{5.3}$$

$$C(t) = C(0)e^{-(L/V)t}$$

We might also ask what happens if there is a steady input I' before time zero that changes to a new rate I at time zero. We assume that before time zero, I' has existed long enough for the system to reach a steady state in which

$$C'_{ss} = I'/L$$

We then use this value C'_{ss} as the zero time concentration in the Laplace transform solution for time greater than zero:

$$\frac{dC(t)}{dt} = \frac{I}{V} - \frac{L}{V} C(t), \qquad t > 0$$

$$sg(s) - \frac{I'}{L} = \frac{I}{sV} - \frac{L}{V} g(s)$$

$$g(s) = \frac{I}{Vs\left(s + \frac{L}{V}\right)} + \frac{I'}{L\left(s + \frac{L}{V}\right)}$$

$$C(t) = \frac{I}{L}\left(1 - e^{-(L/V)t}\right) + \frac{I'}{L} e^{-(L/V)t} \tag{5.4}$$

3. The Two-Compartment Problem

Now let us proceed to a two-compartment problem. To help in visualizing this more complicated case, we adopt the following conventions. An arrow drawn down from the top represents a concentration-independent input to the compartment, or by choosing I negative, a concentration-independent loss from the compartment. An arrow below represents a concentration-dependent loss and an arrow between compartments represents diffusion between the compartments. Diffusion is normally concentration dependent. If, however, there exists active transport between the compartments, it can be incorporated (from left to right) as a negative quantity in I_1 and an equivalent positive quantity in I_2. We diagram a two-compartment system as in Figure 5-1, where L_1C_1 represents concentration-dependent loss from V_1, L_2C_2

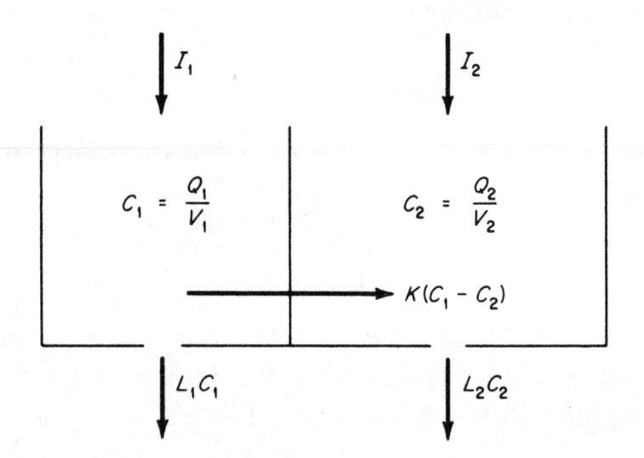

Figure 5-1. A representation of the general two-compartment problem in which concentration-dependent loss is represented by holes through the bottom of the compartments. Concentration-independent loss, or gain, is indicated by the arrows from above, and diffusion between the compartments by the arrows through the barrier between them.

represents concentration-dependent loss from V_2, and $K(C_1 - C_2)$ represents diffusion between the two compartments. Note that V_1 and V_2 are assumed constant. As usual, we write the differential equations describing this system in terms of changes in the quantities Q_1 and Q_2 which occur during a small interval of time Δt.

$$\begin{aligned} \Delta Q_1 &= \Delta t(I_1 - L_1C_1 - K(C_1 - C_2)) \\ \Delta Q_2 &= \Delta t(I_2 - L_2C_2 - K(C_2 - C_1)) \end{aligned} \tag{5.5}$$

(The reader should check the dimensions of L and K.) These changes in quantities are converted into changes in concentration by dividing each equation by the volume of the appropriate compartment

$$\Delta C_1 = \frac{\Delta t}{V_1} [I_1 - L_1 C_1 - K(C_1 - C_2)]$$

$$\Delta C_2 = \frac{\Delta t}{V_2} [I_2 - L_2 C_2 - K(C_2 - C_1)]$$

These, in turn, are converted to differential equations by dividing by Δt and taking the limit as Δt goes to zero:

$$\frac{dC_1}{dt} = \frac{1}{V_1} [I_1 - L_1 C_1 - K(C_1 - C_2)] \tag{5.6a}$$

$$\frac{dC_2}{dt} = \frac{1}{V_2} [I_2 - L_2 C_2 - K(C_2 - C_1)] \tag{5.6b}$$

The steady-state solution is easily found by setting both derivatives equal to zero and solving for C_1 and C_2:

$$C_{1\,ss} = \frac{I_1 L_2 + K(I_1 + I_2)}{L_1 L_2 + K(L_1 + L_2)}$$

$$C_{2\,ss} = \frac{I_2 L_1 + K(I_1 + I_2)}{L_1 L_2 + K(L_1 + L_2)} \tag{5.7}$$

We now wish to investigate the time-dependent solutions in which C_1 and C_2 are initially zero and an input begins at time zero. In the general case, time-dependent solutions are quite complicated. Not only are they tedious to find; they are hard to understand as well. When confronted with problems of this type, the skilled mathematician usually looks for special circumstances that yield simpler solutions. These are often limiting cases in which certain parameters are so small that they can be neglected or so large that other parameters of the problem can be neglected relative to the large ones.

In the two-compartment problem there are a number of such possibilities. We can let one of the concentration-independent inputs be zero; we can let the diffusion constant be very large or very small; or we can look at the cases in which one compartment is much larger than the other. Each of these will be substantially easier then the general case. Sometimes it pays to solve the general case and reduce it to each of the special cases at the end. Other times it is easier to solve the special cases separately. We will initially try the latter method.

Let us consider the case where I_2, $C_1(0)$, and $C_2(0)$ are zero. I_1 starts at zero and has a fixed value thereafter. Diffusion into the second compartment does not significantly affect the rate of buildup in the first. This last statement could be true in two ways. The diffusion constant K can be very small, or the volume of the second compartment can be small and its concentration-dependent loss rate small relative to that of the first compartment. In these special circumstances the diffusion term of Eq. (5.6a) can be neglected in solving for $C_1(t)$. When this is done, (5.6a) reduces to the familiar form

$$\frac{dC_1(t)}{dt} + \frac{L_1}{V_1} C_1(t) = \frac{I_1}{V_1}, \qquad C_1(0) = 0$$

the solution to which we know to be

$$C_1(t) = \frac{I_1}{L_1}(1 - e^{-(L_1/V_1)t}) \tag{5.8}$$

We can now use this value of C_1 in Eq. (5.6b) to find an approximate value for C_2, which we do by means of the Laplace transform, noting that $C_2(0) = 0$.

$$\frac{dC_2(t)}{dt} + \frac{L_2 + K}{V_2} C_2(t) = \frac{KI_1}{L_1 V_2}(1 - e^{-(L_1/V_1)t})$$

$$(s + a)g_2 = \frac{KI_1}{L_1 V_2}\left(\frac{1}{s} - \frac{1}{s + b}\right)$$

where

$$a = \frac{K + L_2}{V_2} \qquad \text{and} \qquad b = \frac{L_1}{V_1}$$

Then

$$g_2 = \frac{KI_1}{L_1 V_2}\left[\frac{1}{s(s + a)} - \frac{1}{(s + a)(s + b)}\right]$$

$$= \frac{KI_1}{L_1 V_2}\left(\frac{b}{s(s + a)(s + b)}\right)$$

The inverse of this form is entry 18 in Appendix II, from which we find

$$C_2(t) = \frac{KI_1}{L_1 V_2 a}\left(1 + \frac{be^{-at} - ae^{-bt}}{a - b}\right)$$

$$= \frac{KI_1}{L_1(K + L_2)}\left[\frac{a(1 - e^{-bt}) - b(1 - e^{-at})}{a - b}\right] \tag{5.9}$$

This is still a fairly complicated result. Let us therefore look at two subcases, which are distinguished by whether a is greater than b or b greater than a. If a is much greater than b, that is,

$$\frac{K + L_2}{V_2} \gg \frac{L_1}{V_1}$$

then $C_2(t)$ simplifies to

$$C_2(t) = \frac{KI_1}{L_1(K + L_2)}(1 - e^{-(L_1/V_1)t}) \tag{5.10}$$

Comparing Eq. (5.10) with Eq. (5.8), we note that the maximum value of C_2 (the steady-state value) consists of a term I_1/L_1 which is the steady-state concentration of compartment one multiplied by $K/(L_2 + K)$. The time constant, however, is that of the first compartment. If

$$b \gg a \qquad \text{or} \qquad \frac{L_1}{V_1} \gg \frac{L_2 + K}{V_2}$$

where V_2 is large, for example, then C_2 reduces to

$$C_2(t) = \frac{KI_1}{L_1(K + L_2)}(1 - e^{-[(K+L_2)/V_2]t})$$

In this case, the limiting value has the same form as in the earlier case but the *time constant* is determined by the diffusion constant and the properties of the *second* compartment.

Now let us examine another case in which K is very large. Specifically, let K be much greater than twice the sum of L_1 and L_2. Physically, we would expect this to result in the two compartments' behaving very much as if they were a single compartment, and we shall see that if L_2 is small, this is indeed true.

In order to solve the case of K large, let us proceed with the general solution and at the end insert the appropriate approximations for K large. We begin by taking the Laplace transform of Eqs. (5.6) for the case in which both concentrations are initially zero and I_2 is zero.

$$V_1 s g_1 = \frac{I_1}{s} - L_1 g_1 - K g_1 + K g_2$$

$$V_2 s g_2 = -L_2 g_2 - K g_2 + K g_1$$

It is helpful now to introduce the simplifying notation

$$k_1 = \frac{K}{V_1}, \quad k_2 = \frac{K}{V_2}, \quad l_1 = \frac{L_1 + K}{V_1}, \quad l_2 = \frac{L_2 + K}{V_2}, \quad r_1 = \frac{I_1}{V_1} \tag{5.11}$$

in terms of which

$$g_1(s + l_1) - k_1 g_2 = \frac{r_1}{s}$$

$$-k_2 g_1 + (s + l_2) g_2 = 0$$

This pair of equations is now easily solved for g_1 and g_2:

$$g_1 = \frac{r_1 + r_1 l_2/s}{s^2 + (l_1 + l_2)s + l_1 l_2 - k_1 k_2}$$

$$g_2 = \frac{r_1 k_2/s}{s^2 + (l_1 + l_2)s + l_1 l_2 - k_1 k_2}$$

and these can now be inverted (see Chapter IV, Section 10) to yield

$$C_2(t) = \frac{r_1 k_2}{a - b}\left[\frac{1 - e^{-bt}}{b} - \frac{1 - e^{-at}}{a}\right] \tag{5.12}$$

$$C_1(t) = r_1 \frac{e^{-bt} - e^{-at}}{a - b} + \frac{l_2}{k_2} C_2(t) \tag{5.13}$$

and note that

$$C_1(t) = \frac{l_2}{k_2} C_2(t) + \frac{1}{k_2}\frac{dC_2}{dt} \tag{5.14}$$

where now

$$a = \frac{p}{2}\left(1 + \sqrt{1 - \frac{4q}{p^2}}\right)$$

$$b = \frac{p}{2}\left(1 - \sqrt{1 - \frac{4q}{p^2}}\right)$$

and in terms of the original notation

$$p = l_1 + l_2 = \frac{K + L_1}{V_1} + \frac{K + L_2}{V_2} = a + b$$

$$q = l_1 l_2 - k_1 k_2 = \frac{L_1 L_2 + K(L_1 + L_2)}{V_1 V_2} = ab$$

Thus far this solution is exact. We note that as t becomes very large, the values of the concentration approach the steady-state values for I_2 zero given by (5.7):

$$C_2(\infty) = \frac{r_1 k_2}{ab} = \frac{I_1 K}{L_1 L_2 + K(L_1 + L_2)}$$

$$C_1(\infty) = \frac{l_2}{k_2} C_2(\infty) = \frac{I_1(K + L_2)}{L_1 L_2 + K(L_1 + L_2)}$$

When we apply the approximation of K very large compared to twice the sum of L_1 and L_2, we find

$$p \approx \frac{K}{V_1} + \frac{K}{V_2} = K\frac{V_1 + V_2}{V_1 V_2}, \qquad q = \frac{K(L_1 + L_2)}{V_1 V_2}$$

and therefore

$$\frac{4q}{p^2} = \frac{4K(L_1 + L_2)\,V_1 V_2}{K^2(V_1 + V_2)^2} < \frac{2(L_1 + L_2)}{K} \ll 1$$

since K is greater than twice $L_1 + L_2$ and $(V_1 + V_2)^2$ is greater than $2V_1V_2$, so that we can apply (4.5) to get simple approximate values for a and b:

$$b = \frac{q}{p} = \frac{L_1 + L_2}{V_1 + V_2}$$

$$a = p - b = \frac{K(V_1 + V_2)}{V_1 V_2} - \frac{L_1 + L_2}{V_1 + V_2}$$

$$= \frac{K(V_1 + V_2)^2 - V_1V_2(L_1 + L_2)}{V_1V_2(V_1 + V_2)}$$

Now the square of $V_1 + V_2$ is certainly larger than the product V_1V_2; furthermore, K is larger than $L_1 + L_2$. Therefore to a good approximation for large K

$$a = K\frac{V_1 + V_2}{V_1 V_2} \tag{5.15}$$

Similarly,

$$\frac{a}{b} = \frac{K(V_1 + V_2)^2}{V_1V_2(L_1 + L_2)} \gg 1$$

Thus a is much greater than b. Since a is much greater than b, we can approximate C_1 and C_2 for large values of t by

$$C_2(t) = \frac{r_1 k_2}{ab}(1 - e^{-bt}) = C_2(\infty)(1 - e^{-bt})$$

$$= \frac{I_1}{L_1 + L_2}\{1 - e^{-[(L_1+L_2)/(V_1+V_2)]t}\} \tag{5.16}$$

Applying Eq. (5.14) we find an approximate value for $C_1(t)$

$$C_1(t) = \frac{K + L_2}{K}C_2(t) + \frac{r_1}{a}e^{-bt} \tag{5.17}$$

from which we see that for large values of time the intuitive idea that the two compartments act as if they were a single compartment of volume $V_1 + V_2$ and concentration-dependent loss $(L_1 + L_2)C$ is essentially correct, except that C_1 reaches a slightly higher value than C_2.

For small values of time, the situation is a bit more complicated. Even if K is very large, the second compartment's filling is still dependent upon the buildup of concentration in the first compartment. We therefore expect C_2 to build up slowly at first.

Subtracting C_2 from (5.13) and replacing l_2 and k_2, we find

$$C_1 - C_2 = \frac{L_2}{K} C_2 + r_1 \frac{e^{-bt} - e^{-at}}{a - b} \approx \frac{L_2}{K} C_2 + \frac{r_1}{a} e^{-bt}$$

and since $a \gg b$ from (5.12), we have

$$\frac{C_2(t)}{C_2(\infty)} = 1 - e^{-bt}, \qquad e^{-bt} = 1 - \frac{C_2(t)}{C_2(\infty)}$$

Now L_2/K is small so that

$$C_1(t) - C_2(t) = \frac{r_1}{a} e^{-bt} = \frac{I_1 V_2}{K(V_1 + V_2)} \left(1 - \frac{C_2(t)}{C_2(\infty)}\right)$$

and we find that as long as C_2 is small, C_1 leads C_2 by approximately

$$C_1 - C_2 = \frac{I_1 V_2}{K(V_1 + V_2)} \tag{5.18}$$

Let us review the nature of the solutions of the two-compartment diffusion problem. Except in the case noted later, both compartments' concentrations go approximately as a simple inverted exponential toward their limiting values given by the steady-state solutions (5.7). If the diffusion constant is small relative to L_1 and L_2, the first compartment acts as if it alone were present. The second compartment has approximately the same time constant as the first (not the same limiting value, however) if either its volume is small or its concentration-dependent loss is large. If, however, the compartment is large and L_2 is small, its time constant is set by the properties of the second compartment.

If the diffusion constant is large compared to the sum of L_1 and L_2, the time constant of the second compartment for large values of time acts as if the two compartments were a single compartment with concentration-dependent loss $(L_1 + L_2)C$ and volume $V_1 + V_2$. For small values of time, C_2 is delayed behind C_1, so that at the time zero it is not an inverted exponential but starts with zero slope and then follows C_1 (Figure 5-2) as described in Eq. (5.18).

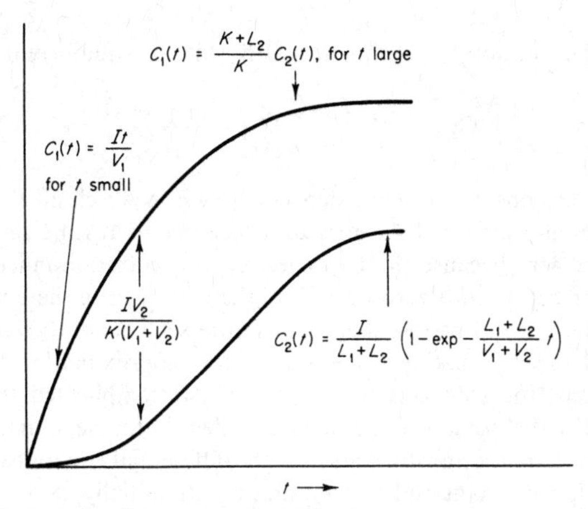

Figure 5-2. Concentration as a function of time in a two-compartment problem with the input only to the first compartment, showing how the concentration in the first compartment leads the concentration in the second compartment.

4. A Simple Three-Compartment Problem

In the previous example we assumed that the input to the first compartment was a given function of time. Now let us consider a slightly different case in which the input is from a fixed concentration through a diffusing barrier. This situation can be represented diagrammatically as in Figure 5-3. In this case, I_3 is usually a negative quantity representing a concentration-independent loss. We write the differential equations describing this system as

$$V_3 \frac{dC_3}{dt} = I - LC_3 + K_{23}(C_2 - C_3)$$
$$V_2 \frac{dC_2}{dt} = K_{12}(C_1 - C_2) - K_{23}(C_2 - C_3) \tag{5.19}$$

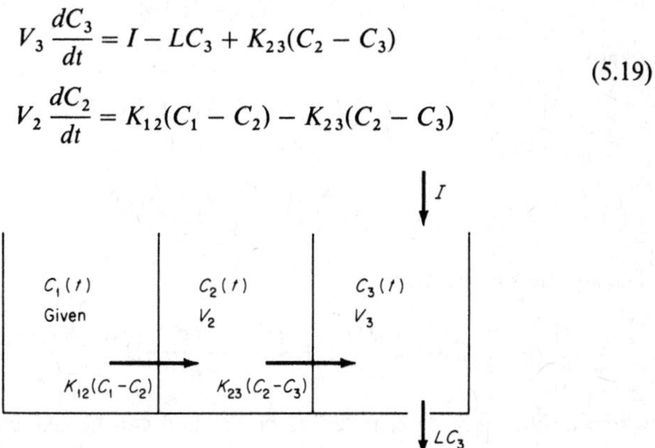

Figure 5-3. A representation of a three-compartment problem.

for which the solution for C_3 in the steady state or equilibrium condition is

$$C_{3\,\mathrm{ss}} = \frac{K_{12} K_{23} C_1 + I(K_{12} + K_{23})}{(K_{12} + K_{23})L + K_{12} K_{23}} \tag{5.20}$$

Now let us consider the time-dependent case in which all the concentrations are initially zero and C_1 goes to a fixed value at time zero. We must exercise care here because the hypothesized concentration-independent loss cannot occur before time zero or shortly thereafter, since there is nothing in compartment three to be lost. Let us therefore set I equal to zero. While we could avoid this by making some appropriate approximation for I at low concentrations, the solutions would then be so complicated that they are better done by the numerical methods described in the next chapter.

Apply the Laplace transformation to the differential equations, remembering that C_1 is a constant and that C_2 and C_3 are initially zero.

$$V_3 s g_3 = -L g_3 + K_{23}(g_2 - g_3)$$

$$V_2 s g_2 = K_{12}\left(\frac{C_1}{s} - g_2\right) - K_{23}(g_2 - g_3)$$

Solve the first of these for g_2

$$g_2 = \frac{g_3}{K_{23}}(V_3 s + L + K_{23})$$

and replace g_2 in the second equation

$$g_3 = \frac{K_{12} K_{23} C_1 / V_2 V_3}{s(s^2 + ps + q)}$$

where

$$p = \frac{L + K_{23}}{V_3} + \frac{K_{12} + K_{23}}{V_2}$$

$$q = \frac{(K_{12} + K_{23})L + K_{12} K_{23}}{V_2 V_3}$$

The solution is therefore

$$C_3(t) = \frac{1}{ab}\left(1 + \frac{ae^{-bt} - be^{-at}}{b - a}\right)\left(\frac{K_{12}K_{23}C_1}{V_2 V_3}\right) \tag{5.21}$$

where a and b are determined as before and can be reduced to manageable forms in limiting cases where L is very large or small, or one compartment is much larger than the others.

5. The Unit Impulse Response

Thus far, with few exceptions, we have considered only very simple inputs to compartmental problems: those that start at time zero and have a fixed value thereafter, and inputs that consist of an initial concentration in one of the compartments. In the real world we are often faced with inputs that vary as functions of time. We have already seen one such case, the input to the second compartment of a two-compartment series dilution, and solved it without great difficulty. However, the input in this case was a very simple function of time, a simple decreasing exponential, and had the input been a more complicated function of time, we would possibly have had difficulty with a brute-force solution.

The unit impulse response provides a way around this kind of difficulty. If, in a linear system, one can define the effect of a unit input at time zero, one can use this result to find the effect of an input that is an arbitrary function of time.

We define the unit impulse response as the ratio of the dependent variable as a function of time to the size of an instantaneous input to the system at time zero. Thus, if in a compartmental problem the concentration $C(t)$ at some point in the system is proportional to the initial injection Q

$$C(t) = QH(t)$$

then $H(t)$ is the unit impulse response.

We have already done the work of calculating H for a number of cases. In the one-compartment dilution we found, for an instantaneous input Q, that

$$C(t) = \frac{Q}{V} e^{-(F/V)t}$$

Thus

$$H(t) = \frac{C(t)}{Q} = \frac{1}{V} e^{-(F/V)t} \tag{5.22}$$

(Note that the units of H are not the same as the units of C.)

We have also done the two-compartment series dilution, where we found for the concentration in the second compartment

$$C_2(t) = \frac{Q}{V_1 - V_2}(e^{-(F/V_1)t} - e^{-(F/V_2)t})$$

Thus

$$H_2(t) = \frac{C_2(t)}{Q} = \frac{1}{V_1 - V_2}(e^{-(F/V_1)t} - e^{-(F/V_2)t}) \tag{5.23}$$

To calculate the unit impulse response one usually uses the Laplace transform and builds the effect of the input Q into the initial conditions. We shall illustrate this again by solving the two-compartment diffusion problem that we discussed earlier in this chapter by an alternative method, in which we eliminate one first-order differential equation at the cost of having to solve, instead, a single second-order differential equation. From (5.6a) and (5.6b) we have

$$\begin{aligned} \frac{dC_1}{dt} + \frac{L_1 + K}{V_1} C_1 - \frac{K}{V_1} C_2 &= \frac{I_1(t)}{V_1} \\ \frac{dC_2}{dt} + \frac{L_2 + K}{V_2} C_2 - \frac{K}{V_2} C_1 &= \frac{I_2(t)}{V_2} \end{aligned} \tag{5.24}$$

To find the effect of a unit input we are going to let I_1 and I_2 be zero and incorporate the effect of an impulse Q into the initial conditions

$$C_1(0) = \frac{Q}{V_1}, \qquad C_2(0) = 0$$

$$\frac{dC_2(0)}{dt} = +\frac{K}{V_2} C_1(0) = +\frac{KQ}{V_1 V_2}$$

Let us introduce the compact symbols we used before (5.11):

$$\frac{dC_1}{dt} + l_1 C_1 - k_1 C_2 = 0, \qquad \frac{dC_2}{dt} + l_2 C_2 - k_2 C_1 = 0$$

Now we eliminate the variable C_1 by solving the second equation for C_1 in terms of C_2, differentiating it, and inserting the results into the first equation of the foregoing pair.

$$\begin{aligned} C_1 &= \frac{1}{k_2}\left(\frac{dC_2}{dt} + l_2 C_2\right) \\ \frac{dC_1}{dt} &= \frac{1}{k_2}\left(\frac{d^2C_2}{dt^2} + l_2 \frac{dC_2}{dt}\right) \\ \frac{d^2C_2}{dt^2} + (l_2 + l_1)&\frac{dC_2}{dt} + (l_1 l_2 - k_1 k_2) C_2 = 0 \end{aligned} \tag{5.25}$$

We can now solve this by means of the Laplace transform, noting that we know the values of $C_2(0)$ and $dC_2(0)/dt$:

$$s^2 g - sC_2(0) - \frac{dC_2(0)}{dt} + (l_1 + l_2)(sg - C_2(0)) + (l_1 l_2 - k_1 k_2) g = 0$$

$$g = \frac{KQ}{V_1 V_2} \frac{1}{s^2 + (l_1 + l_2)s + (l_1 l_2 - k_1 k_2)}$$

Thus the result of an impulse Q at time zero is

$$C_2(t) = \frac{e^{-at} - e^{-bt}}{b - a} \frac{KQ}{V_1 V_2} \tag{5.26}$$

where

$$a = \frac{p}{2}\left(1 + \sqrt{1 - \frac{4q}{p^2}}\right)$$

$$b = \frac{p}{2}\left(1 - \sqrt{1 - \frac{4q}{p^2}}\right)$$

$$p = l_1 + l_2$$

$$q = l_1 l_2 - k_1 k_2$$

and the unit impulse response is

$$H(t) = \frac{K}{V_1 V_2} \frac{e^{-at} - e^{-bt}}{b - a} \tag{5.27}$$

We shall now show how the concept of the unit impulse response can be used to find the effect of an input that is an arbitrary function of time.

6. The Convolution Theorem

Let us return to the simple one-compartment problem for the concentration-dependent case; specifically, let us consider that the concentration change is due to flow through the compartment.

$$\frac{dC}{dt} = \frac{1}{V}(I(t) - FC)$$

We know that for an impulse injection of quantity Q_0 at time zero we have the simple result

$$C(t) = C(0)e^{-(F/V)t} = \frac{Q_0}{V} e^{-(F/V)t}$$

Consider what would happen if we have two impulse injections, one of Q_0 occurring at time zero followed by a second injection at, say, time 3 sec, of amplitude Q_3. Intuitively we feel that the result of two such injections might simply be the sum of the resulting concentrations, which would be due to the injections had they occurred separately. But let us see whether or not we can rigorously establish this principle, which is called the *principle of linear superposition*. Consider first the result of the injection that occurs at time zero.

Clearly, up until the time of the second injection at 3 sec, the concentration will be identical to that given above

$$C(t) = C(0)e^{-(F/V)t} = \frac{Q_0}{V}e^{-(F/V)t}$$

At 3 sec, however, the concentration will change abruptly due to the second injection and it will change in the amount equal to the quantity of tracer injected at 3 sec divided by the volume. But to this must be added the concentration that remains at time 3 sec from the initial injection; therefore, just after the impulse injection at 3 sec, the concentration will be given by*

$$C(3) = \frac{Q_0}{V}e^{-(F/V)3} + \frac{Q_3}{V}$$

if F is in units of volume per second. For times greater than 3 sec we can think of this as a new problem, a simple dilution beginning at 3 sec with a concentration $C(3)$.

Thus, for $t > 3$ sec,

$$C(t) = C(3)e^{-(F/V)(t-3)}$$

Note that the exponent contains $t - 3$, so that the exponential factor is one at 3 sec. If we write the results in the following form, we see that after 3 sec, the result of two impulse injections is the sum of the concentrations that would have occurred had either injection been given alone.

$$C(t) = \left(\frac{Q_0}{V}e^{-(F/V)3} + \frac{Q_3}{V}\right)e^{-(F/V)(t-3)}, \qquad t > 3$$

$$= \frac{Q_0}{V}e^{-(F/V)t} + \frac{Q_3}{V}e^{-(F/V)(t-3)}, \qquad t > 3$$

Now, clearly, if this result applies to the case of two input injections, it will apply to many; it is a general principle for systems of linear differential equations that the result of many inputs is identical to the sum of the results that each would have produced independently.

In order to extend this to an arbitrary input as a function of time, we retreat slightly from the concept of an impulse injection of quantity Q_0 to an injection of the same quantity of tracer over a very small, but nonzero, interval of time Δt. If Δt is sufficiently small, the result of an injection during this time will be indistinguishable from a true impulse except during the time Δt.

Now let us consider a function of time $I(t)$ that is zero before time zero but arbitrary thereafter, and let us break this up into a number of small time intervals of duration Δt each (Figure 5-4), so that

$$Q(n) = \Delta t \, I(n \, \Delta t)$$

*For clarity, dimensions are left out in the next few lines.

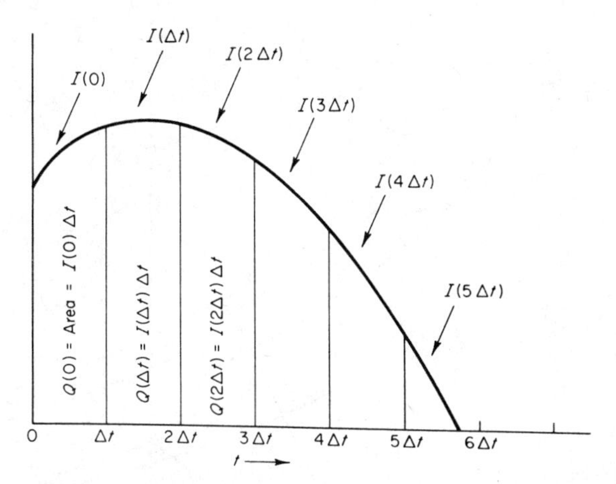

Figure 5-4. How a continuous input is described as a function of time in terms of infinitesimal units of time. Within each unit of time Δt, a quantity of tracer ΔQ equal to $I(t)$ Δt enters the system.

By means of the linear superposition principle, we could calculate the effect of each little rectangle of input shown in Figure 5-4 and add up the resulting concentrations as in Figure 5-5. Let us do so at some fixed time T. We note that the contribution to the concentration at time T due to the little rectangle of input $Q(0)$ is

$$C = \frac{Q(0)}{V} e^{-(F/V)T}$$

while the concentration due to the rectangle $Q(\Delta t)$ is

$$C = \frac{Q(\Delta t)}{V} e^{-(F/V)(T-\Delta t)}$$

where we take particular note of the Δt that occurs in the exponent. In the preceding calculation the exponent contained only the real time T, but in this case the time that $Q(\Delta t)$ had to decay is not the full interval of time from zero to T but the difference between T and the time at which this rectangle begins, which is Δt. Similarly, when we calculate the result of the rectangle that starts at $2\,\Delta t$, we must consider that this has had a time to decay $T - 2\,\Delta t$. We may therefore write the total concentration at time T as

$$C(T) = \frac{1}{V}[Q(0)e^{-(F/V)T} + Q(\Delta t)e^{-(F/V)(T-\Delta t)} + Q(2\,\Delta t)e^{-(F/V)(T-2\,\Delta t)}$$
$$+ \cdots + Q(6\,\Delta t)e^{-(F/V)(T-6\,\Delta t)}] \qquad (5.28)$$

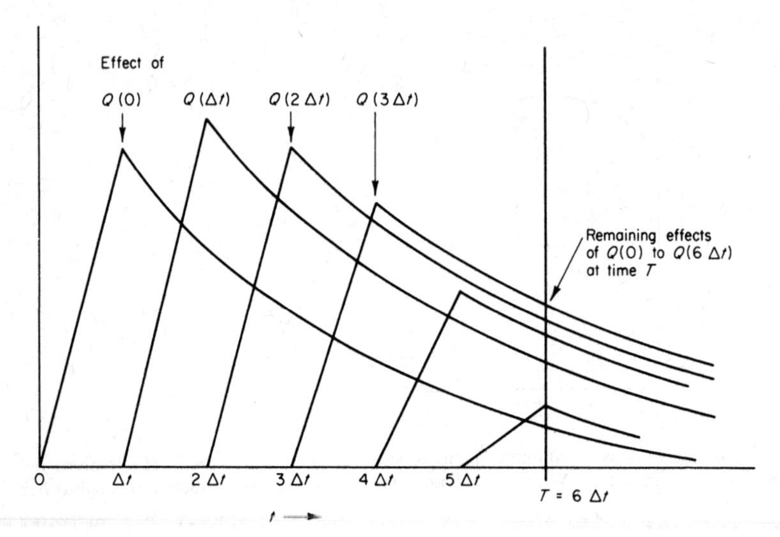

Figure 5-5. The effect of concentration measured at time 6 Δt as a function of inputs that have occurred in each of the infinitesimal units of time Δt.

We can write this in terms of the rate of input I as

$$C(T) = \frac{\Delta t}{V}\left[I(0)e^{-(F/V)T} + I(\Delta t)e^{-(F/V)(T-\Delta t)} + \cdots + I(6\,\Delta t)e^{-(F/V)(T-6\,\Delta t)}\right] \tag{5.29}$$

Now look closely at the nature of each of the terms in this series. Each term consists of the product of the input, which occurred at some time prior to T, multiplied by a term that represents how much this input has decayed between the time of its injection and T. Let us make a plot of each of these terms, not as a function of T, the time at which we want to know the concentration, but rather as a function of the time of injection of each little rectangular area (Figure 5-6). Examine this logic carefully; it is a bit tricky. Now replace the approximate areas represented by the sum by the exact areas. For each fixed value of T, we plot the effect due to each injection that occurs prior to T, and thus the concentration at time T is the area under the curve defined by the rectangles in Figure 5-6, which we write as a definite integral in the form

$$C(T) = \int_0^T I(\tau)\,\frac{1}{V}\,e^{-(F/V)(T-\tau)}\,d\tau \tag{5.30}$$

To assure ourselves that this is correct, we can evaluate the quantity inside the

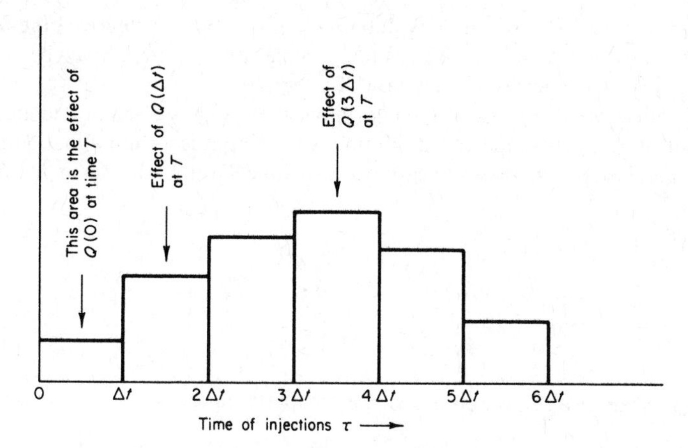

Figure 5-6. A different representation of the effect at time 6 Δt in terms of the quantity of input which occurs in each infinitesimal unit of time prior to 6 Δt.

integral at some fixed time $\tau = 3\,\Delta t$. The quantity inside the integral will be

$$I(3\,\Delta t)\frac{1}{V}\,e^{-(F/V)(T-3\,\Delta t)}$$

which is seen to correspond to the fourth term in the series (5.29). Now look at the two terms that make up the product inside the integral. The first of these is the rate of injection as a function of time, but note: not as a function of the time of observation T, but as a function of a new quantity τ, which represents the position along the horizontal axis in Figure 5-6. The second term of the product is the response that occurs at time T due to an input at τ. It is exactly the response due to a unit impulse, except that the time is the observation time minus the time of occurrence of the impulse. The integral (5.30) is called the convolution of I with the impulse response

$$H(t) = \frac{1}{V}\,e^{-(F/V)t}$$

The foregoing derivation did not depend upon the specific form of the unit impulse response. Thus, for any linear system for which a unit impulse can be defined and whose input is zero before time zero we can find the result of an arbitrary input by calculating the convolution of the input with the unit impulse response:

$$C(t) = \int_0^t I(\tau)H(t-\tau)\,d\tau = \int_0^t I(t-\tau)H(\tau)\,d\tau \tag{5.31}$$

Note that two forms of the convolution are given. We have derived the first. The derivation of the second is along similar lines. Either may be used, depending upon their respective ease of integration.

To illustrate the utility of the unit impulse response we shall calculate the output of a single-compartment dilution whose input is a ramp function that starts at zero and increases linearly to I_1 at time T_1 and remains at I_1 thereafter.

Let

$$I(t) = \frac{I_1}{T_1} t, \qquad 0 < t < T_1$$

$$= I_1, \qquad t > T_1$$

The unit impulse response for a single compartment is

$$H(t) = \frac{1}{V} e^{-(F/V)t}$$

To find the convolution for t less than T_1 we replace t in the impulse response by $t - \tau$ and in the input function by τ. Thus

$$C(t) = \int_0^t \frac{I_1}{T_1} \tau \frac{1}{V} e^{-(F/V)(t-\tau)} \, d\tau$$

$$= \frac{I_1}{T_1 V} e^{-(F/V)t} \int_0^t \tau e^{(F/V)\tau} \, d\tau$$

$$= \frac{I_1}{T_1 V} e^{-(F/V)t} \left[\frac{1}{(F/V)^2} e^{(F/V)\tau} \left(\frac{F\tau}{V} - 1 \right) \Big|_0^t \right]$$

$$= \frac{I_1 V}{T_1 F^2} \left[e^{-(F/V)t} + \frac{Ft}{V} - 1 \right]$$

For times greater than T_1 the integral must be split into two parts, since the form of the input function changes. For t greater than T_1,

$$C(t) = \int_0^{T_1} \frac{I_1}{T_1} \tau H(t - \tau) \, d\tau + \int_{T_1}^t I_1 H(t - \tau) \, d\tau$$

$$= \frac{I_1}{F} + \frac{I_1 V}{T_1 F^2} \left[1 - e^{(F/V)T_1} \right] e^{-(F/V)t}$$

We could of course have solved this problem by brute force. We could have solved

$$\frac{dC}{dt} + \frac{F}{V}C = \frac{I_1}{T_1 V}t, \qquad 0 < t < T_1, \quad C(0) = 0$$

$$\frac{dC'}{dt} + \frac{F}{V}C' = \frac{I_1}{V}, \qquad t > T_1, \qquad C'(T_1) = C(T_1)$$

but having once found the unit impulse response, we have already done most of the work and the solution is reduced to the evaluation of a pair of integrals.

There is another very useful relationship involving the unit impulse response and step inputs. If the input to a system is a step from zero to I at time zero and the unit impulse response is $H(t)$, then

$$C(t) = I\int_0^t H(\tau)\,d\tau \tag{5.32}$$

which we obtain by noting that $I(t - \tau)$ is a constant if $t - \tau$ is greater than 0, which it is in the range of the variable τ from 0 to t.

Let us apply this to the unit impulse response for the two-compartment diffusion problem and see if we get the same result as was previously found. From (5.27) we have

$$H(t) = \frac{K}{V_1 V_2}\frac{e^{-at} - e^{-bt}}{b - a}$$

for a step input I.

$$C(t) = \frac{IK}{V_1 V_2}\frac{1}{b - a}\int_0^t (e^{-a\tau} - e^{-b\tau})\,d\tau$$

$$= \frac{KI}{V_1 V_2}\frac{1}{b - a}\left(\frac{1 - e^{-at}}{a} - \frac{1 - e^{-bt}}{b}\right)$$

which is the same as Eq. (5.12). Similarly, for the two-compartment series dilution problem we have solved the equation

$$V_2\frac{dC_2}{dt} + FC_2 = FC_1(0)e^{-(F/V_1)t}, \qquad C_2(0) = 0 \tag{5.33}$$

by two fairly complicated methods. Now let us solve it by a simple method. We know that the unit impulse response of the second compartment is

$$H(t) = \frac{1}{V_2}e^{-(F/V_2)t}$$

Now note that the right side of (5.33) is a rate of input in moles per second. Therefore, let

$$I(t) = FC_1(0)e^{-(F/V_1)t}$$

then

$$C_2(t) = \int_0^t FC_1(0)e^{-(F/V_1)\tau} \frac{1}{V_2} e^{-(F/V_2)(t-\tau)}\, d\tau$$

$$= \frac{FC_1(0)}{V_2} e^{-(F/V_2)t} \int_0^t e^{-F(1/V_1 - 1/V_2)\tau}\, d\tau$$

$$= \frac{V_1 C_1(0)}{V_1 - V_2}(e^{-(F/V_1)t} - e^{-(F/V_2)t})$$

We see that we obtain the same result that we found previously for the concentration of the second compartment but in a much less laborious fashion. Furthermore, this tells us how to do a three-compartment series problem. We can regard the third compartment as a separate problem whose unit impulse response is known to be

$$H(t) = \frac{1}{V_3} e^{-(F/V_3)t}$$

and whose input is

$$I(t) = FC_2(t)$$

Thus

$$C_3(t) = \frac{FV_1C_1(0)}{(V_1 - V_2)V_3} \int_0^t e^{-(F/V_3)(t-\tau)} (e^{-(F/V_1)\tau} - e^{-(F/V_2)\tau})\, d\tau$$

which we leave for the reader to evaluate.

Now, as one might suspect, the impulse response of a linear system has a unique role in the description of the system, which may be stated as follows. Given the differential equation or system of differential equations that characterize the system, these determine an impulse response, and from this impulse response one can calculate the response of the system to an arbitrary input as a function of time. We should note that the converse is not true. A particular impulse response, although it characterizes the system, does not allow us to find a unique corresponding differential equation. It may allow us to find some differential equation that will produce this impulse response, but not a unique one. There will be others whose impulse responses are indistinguishably different from the observed one. Furthermore, given the output of the system to a known input as a function of time, it is not possible uniquely to define the impulse function. Or, given the output and the impulse response, it is not uniquely possible to define the input. However, as we will see later, there are useful approximations to this.

7. Numerical Solutions from the Unit Impulse Response

In the problem above we had the unit impulse response and input in algebraic form. Frequently we have them only numerically, from having injected a bolus of tracer and measured the impulse response of the system. We now want to know its response to a complicated input.

We derived the convolution by replacing a series by an integral. Let us return to the series form. For a single compartment at T equal to 6 Δt we had (5.28). We noted that each term of (5.28) represented the input that occurred in a particular Δt multiplied by a fraction that represented the fraction of the input left at time T.

We then generalized this for an arbitrary unit impulse response and found

$$C(T) = \Delta t[I(0)H(T) + I(\Delta t)H(T - \Delta t) + I(2\Delta t)H(T - 2\,\Delta t) + \cdots + I(T)H(0)]$$

Note that in each product term the sum of the times is T. The sum includes all the time steps from 0 to T in I and the reverse sequence in H. Using this rule we can write

$$\begin{aligned}
C(0) &= \Delta t[I(0)H(0)] \\
C(1) &= \Delta t[I(1)H(0) + I(0)H(1)] \\
C(2) &= \Delta t[I(2)H(0) + I(1)H(1) + I(0)H(2)] \\
&\vdots \\
C(N) &= \Delta t[I(N)H(0) + I(N-1)H(1) + \cdots + I(1)H(N-1) + I(0)H(N)]
\end{aligned} \tag{5.34}$$

Example. Let us try an example. For the unit impulse response, we shall assume a linear decay from the value 1 at time zero to the value 0 at 10 sec. For an input function, we shall use a constant for 10 sec and zero thereafter. We shall carry out the calculations for 1-sec intervals.

$$\begin{aligned}
C(0) &= I(0)H(0) = 10 \times 1 = 10 \\
C(1) &= I(1)H(0) + I(0)H(1) = 10 \times 1 + 10 \times 0.9 = 19 \\
&\vdots \\
C(10) &= I(10)H(0) + I(9)H(1) \cdots I(0)H(10) \\
&= 10 \times 1 + 10 \times 0.9 + \cdots + 10 \times 0 = 55 \\
C(11) &= I(11)H(0) + I(10)H(1) + \cdots + I(0)H(11) \\
&= 0 \times 1 + 10 \times 0.9 + \cdots + 10 \times 0.1 + 0 \times 0 = 45
\end{aligned}$$

The result is plotted in Figure 5-7.

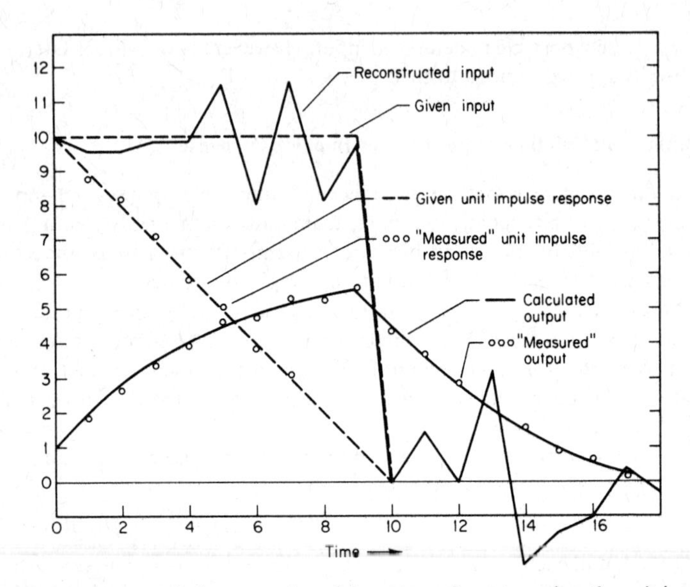

Figure 5-7. A numerical computation of the output of a system when the unit impulse response is known and the input is known. In this case, the unit impulse response is represented by the slanting broken line, while the given input is represented by the rectangular broken line. The output, as calculated from the input and the unit impulse response, is shown as a solid line. For both the unit impulse response and the output, the values were perturbed by a coin flip, yielding the values shown by the dots adjacent to the unit impulse line and the output curve. These perturbed values were then used to reconstruct the input from a unit impulse response and output. The result is the jagged solid line in the figure.

8. Working Backward

Another interesting thing we can do is to work backward. Given the impulse response and the output, it is possible to reconstruct the input. All we have to do is solve Eq. (5.34) for the inputs, in terms of the unit impulse response and output.

$$I(0) = \frac{C(0)/\Delta t}{H(0)}$$

$$I(1) = \frac{C(1)/\Delta t - I(0)H(1)}{H(0)}$$

$$I(2) = \frac{C(2)/\Delta t - I(0)H(2) - I(1)H(1)}{H(0)}$$

$$\vdots$$

If we insert the outputs from our previous calculation, we should get back the correct inputs. This process is often used by pharmacologists to find the input to a compartment where the output is known and the impulse function can be measured by injecting a labeled compound. Unfortunately, the technique is not without difficulty. Adjacent to the unit impulse response and the calculated output in Figure 5-7 are a series of dots. These were obtained by changing each value by 2%, the change being positive or negative, depending on the flip of a coin. In this way, we have simulated small experimental errors that would occur if one were really trying to carry out this process. We have then taken the disturbed values and calculated the input. The result is the highly jagged line shown in the figure. The difficulty, of course, is that errors are cumulative. Input at time zero is fairly accurately calculated, but the calculation of input at time 1 depends upon the result of the calculation at time zero. Time 2, in turn, depends on 1 and zero. Thus, small errors tend to build up, and the process rapidly losses its accuracy. On the other hand, it is clearly not useless; in fact, if the experiment were repeated many times and the resulting inputs averaged, one might expect that the jaggedness would not be quite so apparent in the average.

It is interesting to note that one of the earliest applications of the input reconstruction technique discussed above was the correction of errors in pressure measurements due to the response of catheters. It is possible to measure the unit impulse response of a catheter by applying an instantaneous impulse of pressure at its end. The measured pressure as the other end of the catheter is then a unit impulse response. Experimental pressure measurements are often distorted by the unit impulse response of the catheter; by this technique, however, one can work backward and find the input from the measured output and the unit impulse response.

Problems

1. Consider a cell that maintains its internal concentration C_{in} of some substance below that of the fixed external environment C_{ext} by means of a pump, which pumps at a fixed rate of R moles/sec against a diffusion input of K moles/sec/mole/liter concentration difference. Find the steady state value of C_{in} in terms of R, K, and C_{ext}. If the pump stops operating at time zero, describe the concentration C_{in} thereafter.

2. Work Problem 1 for the case in which the concentration of the environment is influenced by the existence of the pump. This problem must be considered as a two-compartment problem with a finite volume for the

outer compartment. Problem 1 could have been considered as a two-compartment problem with an infinite outer volume, although this is not the easiest way to solve it.

3. Find the output of a one-compartment dilution when the *input* is zero before time zero and

$$I(t) = I_0 e^{-(F/V)t}$$

after time zero.

4. Suppose one were doing a "stop-flow" experiment in which an organ and its contained blood are brought to a given initial concentration of a tracer C_0. At time zero, blood begins to flow. Describe the concentration in the tissue as a function of time for two cases:

Case 1. Blood and tissue are regarded as separate but each well-mixed compartments. Show that

$$V_1 V_2 \frac{d^2 C_2}{dt^2} + [V_1 K + (F + K) V_2] \frac{dC_2}{dt} + FKC_2 = 0$$

where V_1 is the blood volume, V_2 tissue volume, and K a diffusion constant. Find $dC_2(0)/dt$. Solve this system by the Laplace transform method.

Case 2. Blood compartment is not well mixed but effluent blood has the same concentration of tracer as the tissue.

5. Given a linear system whose unit impulse response is a rectangle plotted against time of duration 1 sec. Find the output for an input that is zero before time zero, I_0 after time zero to 10 sec, and zero thereafter.

$$\begin{aligned} H(t) &= 0, \quad && t < 0 \\ &= 1, && 0 < t < 1 \\ &= 0, && t > 1 \\ I(t) &= 0, && t < 0 \\ &= I_0, && 0 < t < 10 \\ &= 0, && t > 10 \end{aligned}$$

6. In order to illustrate the importance of simplifying cases, consider a two-compartment system as shown in Figure 5-1. At time zero

$$C_1(0) = C_2(0) = 0$$

and I_1 begins. Solve for approximate values of $C_1(t)$ and $C_2(t)$ under the assumption that L_2 is much larger than $K + L_1$. The simplification involved is that when L_2 is much larger than $K + L_1$, C_2 will be small compared to C_1. One can then find a simple form for $C_1(t)$ and use this to calculate $C_2(t)$.

7. A system with a unit impulse response of

$$H(0) = 0,$$
$$H(1) = 0,$$
$$H(2) = 4,$$
$$H(3) = 2,$$
$$H(n) = 0, \qquad n > 3$$

has an input

$$I(0) = 4,$$
$$I(1) = 5,$$
$$I(n) = 0, \qquad n > 1$$

Calculate its output as a function of time.

8. Show that for large values of time, Eqs. (5.12) and (5.13) reduce to the form given by Eq. (5.7).

VI

Numerical Methods

Thus far all of the problems in this book have been solved in what is called analytic form, by which is meant a solution in terms of relatively simple functions that at worst have to be looked up in tables. Many problems that arise in biology are too complicated to do this way, and when these are encountered, one solves them numerically. There are also some problems that one could do analytically but that one chooses to do numerically simply as an easy way of tabulating solutions.

There are many techniques for the numerical solution of differential equations and an extensive literature on the subject. Until recently, however, much of this work has been motivated by the need to minimize the amount of manual computation required. Since the advent of high-speed digital computers, the amount of numerical computation is no longer a serious limiting factor for most problems biologists encounter. Simple techniques can be used, and these have the advantage of being quite general and applicable to a wide variety of problems.

The techniques described in this chapter are intended for use on a digital computer. A few years ago it would not necessarily have been sensible to cast the treatment in this form, but in the 1970s there is no excuse for a scientist not to have some familiarity with the use of digital computers, at least in the higher-level languages, such as FORTRAN, APL, ALGOL, or BASIC. The choice among these is a matter of personal preference or depends upon the computing system available. The author's current preference is for APL.

Where it has been necessary to outline computer programs in this and later chapters, these are done in very general terms rather than in terms of a specific language. Considerable thought was given to whether programs should be displayed in one of the well-known languages, such as FORTRAN. It was decided not to do this because not all readers would be familiar with any one such language. In studying this chapter, it is strongly recommended that the outlined programs be converted to a language with which the reader is familiar and that he has available, and that the programs be tried. If it is necessary to learn a language to do this, be assured that it is time well invested.

1. Euler's Method

To illustrate, we shall begin with some differential equations that we have already solved in analytic form and show how they can be done numerically. For the first example, we use the concentration in the second compartment of the two-compartment series dilution problem. The method used, sometimes called *Euler's method*, is one of conceptual simplicity and great generality.

Given a differential equation

$$\frac{dy}{dx} = f(x, y) \tag{6.1}$$

where the right-hand side is some specific function of y or x or both of these, we insert for y and x their known initial values x_0 and y_0 and use these to compute the small change Δy_1 that will occur in y due to a small change in x. We call the latter a *step* in x.

$$\frac{dy}{dx} = f(x_0, y_0) \tag{6.2}$$

$$\Delta y_1 \approx \Delta x \, f(x_0, y_0) \tag{6.3}$$

We then compute new values for x and y

$$x_1 = x_0 + \Delta x,$$

$$y_1 = y_0 + \Delta y_1$$

and insert these into (6.1) and repeat the process, calculating

$$\Delta y_2 \approx \Delta x \, f(x_1, y_1)$$

$$y_2 = y_1 + \Delta y_2$$

$$x_2 = x_1 + \Delta x$$

To be specific, in the case of the two-compartment dilution where, from (2.13), we have

$$\frac{dC_2(t)}{dt} = \frac{F}{V_2}[C_1(0)e^{-(F/V_1)t} - C_2(t)] \tag{6.4}$$

we take as steps

$$\Delta C_2(t) = \Delta t \frac{F}{V_2} [C_1(0)e^{-(F/V_1)t} - C_2(t)] \tag{6.5}$$

and start with a known initial condition that at t zero, $C_{2,0}$ is zero. We then compute the small change $\Delta C_{2,1}$ that occurs during the first step of time Δt, using the value $t = 0$.

$$\begin{aligned} \Delta C_{2,1} &= \Delta t \frac{F}{V_2} [C_1(0)e^0 - 0] \\ C_{2,1} &= C_{2,0} + \Delta C_{2,1} = \Delta C_{2,1} \end{aligned} \tag{6.6}$$

To keep track of time we set

$$t_1 = t_0 + \Delta t = \Delta t$$

These new values of C_2 and t are then inserted on the right-hand side of Eq. (6.5) and the result of the second step in time is calculated.

$$\Delta C_{2,2} = \Delta t \frac{F}{V_2} [C_1(0)e^{-(F/V_1)t_1} - C_{2,1}]$$

$$C_{2,2} = C_{2,1} + \Delta C_{2,2}$$

$$t_2 = t_1 + \Delta t = 2\,\Delta t$$

The process is then repeated for as many steps in time as we desire.

(For sample programs, see Appendix VII.)

2. Verifying the Result

Provided the steps Δt are taken small enough, this technique will almost always yield the correct solution. Unfortunately, there is no a priori way of knowing how small a step to take. There are, however, three very good tests for whether a sufficiently small step has been used.

The simplest test is to change the size of the step and check to see that the same results are obtained. The changed step should always be smaller than the original step, since making the step smaller can only make the result better. Preferably, it should not be related to the original step by some simple factor, such as one half or one tenth, but should be chosen, for example, as the original step divided by the square root of two. This guards against a remote but possible situation in which the result will appear correct for certain specific isolated values of step but not for those in between.

The second test is the reverse computation test, which is analogous to testing whether a boy scout knows how to find his way in the woods by letting him enter, which is no test, and seeing if he can find his way back to the starting point.

The reverse computation test consists of running the calculation forward in time as described above for however long an interval is required. The process is then reversed by changing the time step to a negative value and going back to zero time. The results going in the two directions should agree. If the steps are not small enough, the reverse computation will yield values that are significantly different from the forward computation. Figures 6-1 and 6-2 were made from the two-compartment series dilution calculation with the parameter values

$$V_1 = 10 \text{ liters}, \qquad V_2 = 20 \text{ liters}$$
$$Q = 1 \text{ mole}, \qquad F = 1 \text{ liter/sec}$$

In Figure 6-1, steps of 1 sec were used, and it is clear that the reverse computation differs significantly from the forward computation. The steps were, therefore, too large. Figure 6-2 illustrates the same computation done with steps of 0.1 sec. In this case, the forward and reverse computations agree quite well. For comparison, the forward computation values are given at 10, 20, and 60 sec, as are exact values calculated from Eq. (2.14). Although there are some exotic differential equations for which this test would not be valid, one is unlikely to encounter these in biological problems.

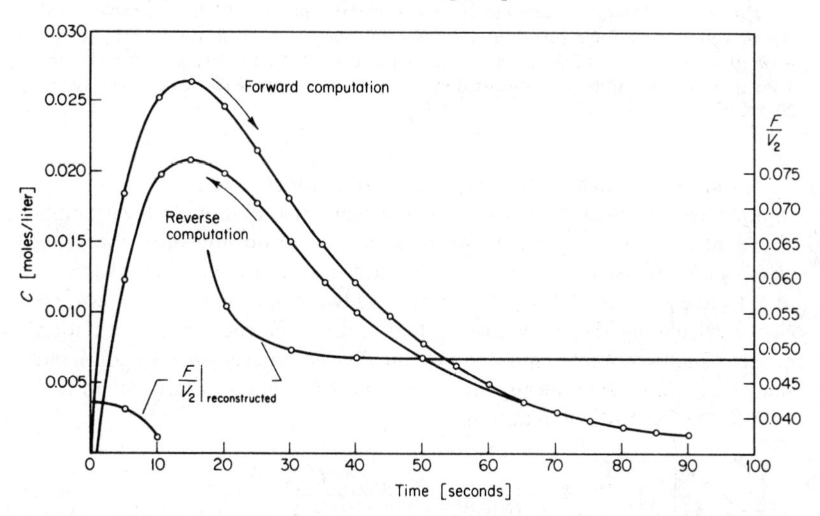

Figure 6-1. Second-compartment concentration of a two-compartment series dilution computed numerically by both forward and reverse computation, showing the disagreement between these two computations when the time interval of computation is taken too large. Also shown, the reconstructed value of F/V_2 from the forward computation.

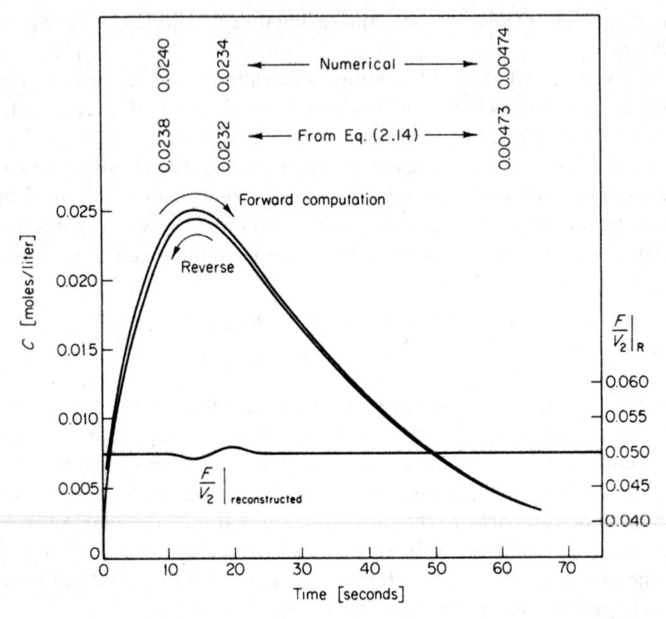

Figure 6-2. Second compartment of a two-compartment series dilution by forward and reverse computation for a smaller time step. Also shown, the reconstructed F/V_2 for the forward computation, and the exact values computed at 10, 20, and 60 sec from Eq. (2.14). Time steps 0.1 sec, except in reconstruction, where 0.01-sec steps were used between 10 and 20 sec.

A third test, which we call the *problem reconstruction* test, starts from the solution and reconstructs the problem to which it is a solution. In the example above, after finding a solution, we pretend that we do not know the value of F/V_2 and we use our calculated concentration curve to compute the value of F/V_2 at each point. This is done by taking two points on the concentration curve, calculating the derivative by finding the difference in concentration between the two points, and dividing this by the bracket on the right-hand side of Eq. (6.4), using mean values for t and $C(t)$, and a Δt larger than that used in the forward computation.

$$\left.\frac{F}{V_2}\right|_{\text{reconstructed}} = \frac{(C_{2,n+1} - C_{2,n})/\Delta t}{C_1(0)e^{-(F/V_1)\frac{1}{2}(t_{n+1}+t_n)} - \frac{1}{2}(C_{2,n+1} + C_{2,n})} \tag{6.7}$$

As noted in Figure 6-2, the reconstruction agrees quite well with the given value of F/V_2 except between 10 and 20 sec, where steps of 0.01 sec had to be used in order to get reasonable values for F/V_2.† The result of the reconstruction should be F/V_2 and, in fact, we find the F/V_2 for which our solution is valid.

The reader may have doubts about the validity of this test, since we seem to risk making the same errors in reconstructing the problem as in the attempt to find a solution. The difference is that in the solution of the problem by Euler's method, errors are cumulative: an error in the first step calculation may introduce a larger error in the next step, and so on. Thus, each point is subject to the accumulated errors of all previously calculated points. The reconstruction process makes use only of a point and its neighbor to calculate F/V_2. Errors therefore are not cumulative.

The Euler method has, in the past, been unpopular because the small size of the steps requires extensive computation. However, the availability of digital computers has removed this objection for most practical purposes, and although there exist far more efficient methods of computation, which allow much bigger steps to be taken, the simplicity of the Euler method is very attractive. Moreover, it has the great virtue that it can be readily generalized to much more complicated systems of equations, which is not always true of the more elegant techniques. For the sake of completeness, however, we enumerate some of the other techniques that are used to find numerical solutions to differential equations. There are at least three others in common use: the *modified Euler method*, which is somewhat more powerful than the one just illustrated; the *Runge–Kutta method*, which is also a stepwise solution; and the *method of Picard*, which is quite different in principle. These three methods are discussed in (15).

3. Numerical Solution of a Circular Reaction with Changing Rate Constant

As another example of a numerical solution, let us consider a three-compartment metabolic loop (described in Chapter II). Assuming, as we did in Chapter II, simple second-order reactions, and assuming that the compounds with which A, B, and C react are present in excess, so that their change in concentration can be neglected, we may write the differential equations describing this process as

$$\frac{dA}{dt} = K_3 C - K_1 A, \qquad \frac{dB}{dt} = K_1 A - K_2 B, \qquad \frac{dC}{dt} = K_2 B - K_3 C$$

† Errors in reconstruction are inevitable at points of reversal such as occur at 14 sec in Figure 6-2. These are a result of division by extremely small numbers and do not represent an error in the problem solution.

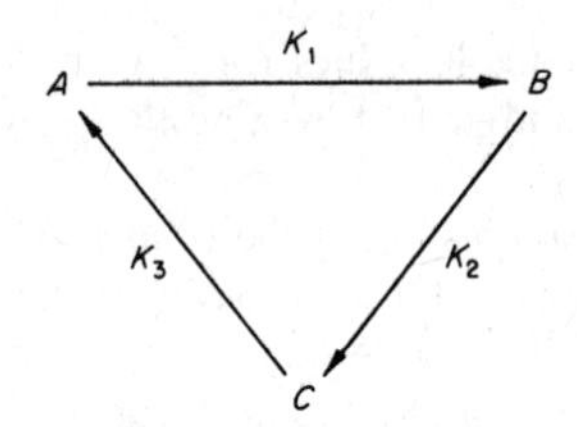

Figure 6-3. A circular reaction with a changing rate constant.

These were solved in Chapter II in the steady state, for which the solutions are

$$A = \frac{Q}{1 + (K_1/K_2) + (K_1/K_3)}, \qquad B = \frac{K_1}{K_2} A, \qquad C = \frac{K_1}{K_3} A$$

where $Q = A + B + C$.

Suppose we ask what would happen if one of the rate constants, K_1, for example, changes with time. It happens that if K_1 changes instantaneously, this system of equations can be solved in closed form. The solution, however, is complicated. The more interesting case is the comparison of the effect of a fast change in the rate constant to the effect of a slow change. The latter case is so difficult analytically that it essentially must be done numerically, so one might as well do both cases numerically. We begin by choosing a set of initial and final values of the rate constants and a total Q of A, B, and C. The starting values of A, B, and C must then satisfy the steady-state conditions for the initial rate constants. We compute the time course of the concentrations by means of an Euler process. For the slow-change case, we allow the rate constant K_1 to be changed during each cycle of the computation until it reaches its final value K_1', where it is fixed for the remainder of the calculation.

1. Set K_1, K_2, K_3 equal to initial values.
2. Set $A = Q \Big/ \left(1 + \dfrac{K_1}{K_2} + \dfrac{K_1}{K_3}\right)$.
3. Set $B = \dfrac{K_1}{K_2} A$.
4. Set $C = \dfrac{K_1}{K_3} A$.
5. Set Time equal to zero.
6. Compute

$$\text{New } A = A + \Delta t(K_3 C - K_1 A)$$

7. Compute

$$\text{New } B = B + \Delta t(K_1 A - K_2 B)$$

8. Compute

$$\text{New } C = C + \Delta t(K_2 B - K_3 C)$$

9. Compute

$$\text{New Time} = \text{Time} + \Delta t$$

10. Check to see if K_1 has reached its maximum allowed value. If it has, skip the next line.
11. Compute

$$\text{New } K_1 = K_1 + \Delta K_1$$

12. Replace old values of K_1, A, B, C, and Time by new values.
13. Print out Time, A, B, and C.
14. Repeat from Step 6.

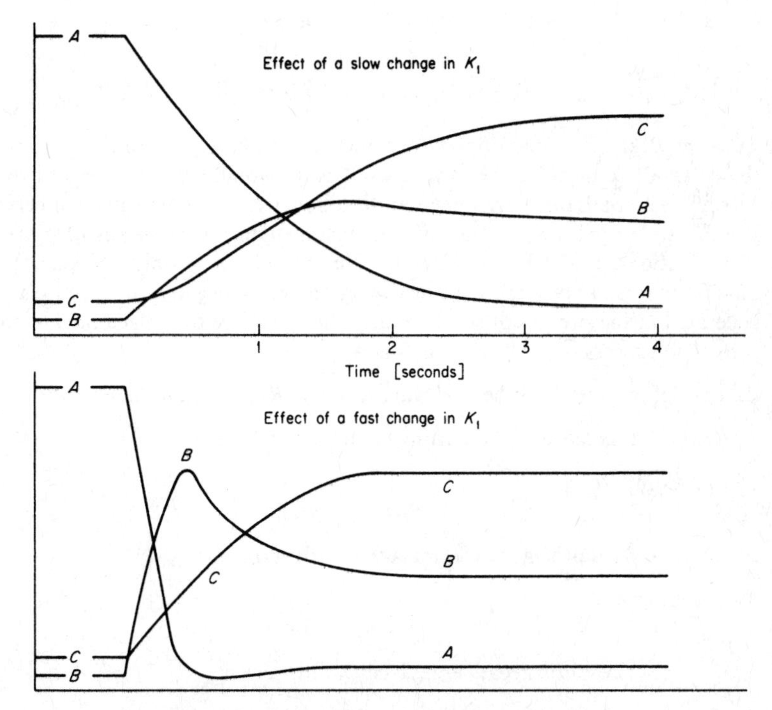

Figure 6-4. Concentrations as a function of time for the circular reaction in the two cases of a slow change in K_1 as opposed to a rapid change in K_1.

The results of such a computation are shown in Figure 6-4 for K_1 starting at 0.1 sec^{-1} and going to 10 sec^{-1} in the interval of time indicated on the graphs. K_2 equals 2 sec^{-1} and K_3 is 1.0 sec^{-1}. Notice in particular the behavior of B in the two cases. In the case of the instantaneous change in K_1, B overshoots its final concentration by about a factor of two. In the case of a slow change, only a very slight overshoot is apparent. A mechanism similar to this, but slightly more complicated, has been suggested as the explanation for the fact that under some conditions a muscle will contract when subject to a rapid depolarization but not to a slow depolarization.

4. Restricted Value Problems

The following equations describe the system shown in Section 9 of the next chapter. If solved analytically, for certain starting conditions, the differential equations will yield negative values for A and B.

$$\frac{dA}{dt} = (R_A - aB) - R_A', \qquad A \geq 0, \quad B \geq 0$$

$$\frac{dB}{dt} = R_B - (R_B' - bA), \qquad R_A - aB \geq 0, \quad R_B' - bA \geq 0$$

A and B represent quantities of substances in metabolic pools and, of course, these cannot be negative. The two quantities in parentheses are rates of irreversible reactions; therefore, they too must be restricted to positive numbers. It is possible to solve problems of this type analytically by means of piecewise solutions, but it is very difficult. Numerically, it is only necessary to modify the usual procedure by checking to see if negative quantities are generated in the process of solution and when they are to replace them with zero. The process of solution is as follows.

1. Set the values of the constants R_A, R_A', R_B, R_B', a, b, Δt.
2. Set A and B to their starting values and t to zero.
3. Compute
$$P = R_A - aB$$
If the result is negative, replace P with zero.
4. Compute
$$Q = R_B' - bA$$
If the result is negative, replace Q with zero.
5. Compute
$$\Delta A = \Delta t(P - R_A')$$

6. Compute

$$\Delta B = \Delta t(R_B - Q)$$

7. Set

$$\text{New } A = A + \Delta A$$

If the result is negative, replace A with zero.

8. Set

$$\text{New } B = B + \Delta B$$

If the result is negative, replace B with zero.

9. Set

$$\text{New } t = t + \Delta t$$

10. Replace t by New t, A by New A and B by New B.
11. Repeat from Step 3.

Figure 6-5 shows the result of one such computation for the set of values

A starting value = 300 μmole	$a = 1\ \text{sec}^{-1}$
B starting value = 200 μmole	$b = 10\ \text{sec}^{-1}$
$R_A = 200$ μmole/sec	$R_A' = 100$ μmole/sec
$R_B = 20$ μmole/sec	$R_B' = 10$ μmole/sec
	$\Delta t = 0.1$ sec

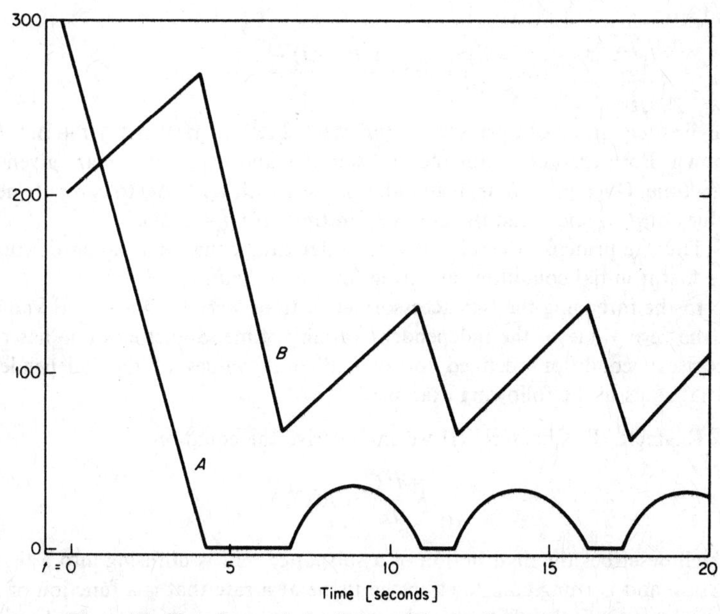

Figure 6-5. Plot of numerical solution of the restricted value problem.

5. Numerical Solution of Higher-Order Equations

The examples given thus far in this chapter have all been first-order differential equations. Let us now see how the method can be extended to second-order equations, which involve a second derivative of the unknown.

$$\frac{d^2y}{dt^2} + p(y, t)\frac{dy}{dt} + q(y, t)y = r(y, t)$$

where p, q, and r are given functions of y and t (given either numerically or analytically). Since this is a second-order equation its solution will contain two arbitrary constants whose values must be determined by accessory conditions of the problem. They can be determined by knowing either the value of y for two values of t or one such value and one for the derivative of y at some value of t. In the latter case the values of both y and its derivative are usually known at time zero, and we shall proceed on that assumption.

In the introduction to Chapter III we showed that the second derivative is given by

$$\frac{d^2y}{dt^2} = \underset{\Delta t \to 0}{\text{limit}} \frac{y(t + 2\,\Delta t) + y(t) - 2y(t + \Delta t)}{(\Delta t)^2}$$

and since

$$\frac{dy}{dt} = \underset{\Delta t \to 0}{\text{limit}} \frac{y(t + \Delta t) - y(t)}{\Delta t}$$

the first equation can be solved for $y(t + 2\,\Delta t)$ if $y(t)$ and $y(t + \Delta t)$ are known. Both are known for the first step if y and its derivative are given at zero time. Once $y(t + 2\,\Delta t)$ is found, it is used with $y(t + \Delta t)$ to compute new values of p, q, and r and the next step in time to $y(t + 3\,\Delta t)$.

Thus, in principle the second-order differential equation is no harder than the first if initial conditions are given for y and dy/dt.

In the foregoing the two accessory conditions were both assumed known at the zero value of the independent variable, time. Sometimes one has the accessory conditions defined for two different values of the independent variable, as in the following example.

Example. In Chapter VIII we shali derive the equation

$$K\frac{d^2C}{dx^2} = R(C)$$

which describes the distribution of a substance that is diffusing into a tissue section and is being consumed in the tissue at a rate that is a function of its concentration. If the relationship between concentration and rate of utilization is a simple one, it is often possible to solve this equation analytically. If, however, $R(C)$ is a complex function, as it often is, it is necessary to solve the differential equation above numerically. Given a tissue slice of thickness H with both of its sides at concentration $C(0)$, we must solve the foregoing with the two boundary conditions $C(0) = C(H)$ given.

The method of solution is to divide the interval 0 to H into N sections Δx numbered 1 through N. Let $C(J)$ be the concentration of the Jth section. The second derivative with respect to x can be approximated as

$$\begin{aligned}\frac{d^2C}{dx^2} &= \frac{1}{\Delta x}\left[\frac{dC(x+\Delta x)}{dx} - \frac{dC(x)}{dx}\right] \\ &= \frac{1}{(\Delta x)^2}\left[(C(x+2\,\Delta x) - C(x+\Delta x)) - (C(x+\Delta x) - C(x))\right] \\ &= \frac{1}{(\Delta x)^2}\left[C(x+2\,\Delta x) + C(x) - 2C(x+\Delta x)\right] \\ &= \frac{R(C)}{K}\end{aligned}$$

Note that $C(x)$ and $C(x + 2\,\Delta x)$ lie on opposite sides of $C(x + \Delta x)$. Thus, in terms of the N sections we compute $C(J)$ in terms of $C(J + 1)$ and $C(J - 1)$ from

$$C(J) = \frac{1}{2}\left(-\frac{R(C)}{K}(\Delta x)^2 + C(J+1) + C(J-1)\right)$$

where this relation must be true for all J except at the end points where

$$C(1) = C(N) \quad \text{given}$$

We will now show how to find a set of values $C(J)$ that satisfy this relation. Within the computer program one provides two tables with N locations each. The first table, which we shall call Table A, is filled with a set of initial values of $C(J)$; we shall explain later how to choose these. The second table, Table B, is generated from the first by the rule

$$C_B(J) = \frac{1}{2}\left(-\frac{(R(C_A(J))}{K}(\Delta x)^2 + C_A(J+1) + C_A(J-1)\right)$$

where $C_A(J)$ means entry number J in Table A and $C_B(J)$ means entry J of Table B and $R(C)$ is calculated from an approximate formula or is found from a table that relates R to C. When all of the entries 1 through N of Table B have been filled, they are transferred to Table A in the corresponding positions except for the two end values, which are set to the end-point boundary conditions. The process is then repeated until a stable set of values is found.

Although in principle this process will converge to the correct values for any reasonable starting values, the convergence can be quite slow. It is therefore helpful to start with values that are known to be close to the correct values. In lieu of a better guess, one can start with zero concentration in the middle and increase it linearly toward the sides, aiming toward the known boundary conditions. However, it is frequently useful to have alternative starting values, as described later, and one alternative would be to start with all the table positions filled with the boundary value.

The process described above is a sequential approximation. It never actually reaches the exactly correct values. One must therefore have some criterion for knowing when to stop. The easiest answer to this is to choose two sets of starting values, one of which is known to be greater than the correct solution at all J and the other known to be less than the correct solution at all J. Thus, one approaches the solution from both sides and traps it between two approximate solutions. One must also be careful that the width of the intervals is small enough so that the result truly approximates a continuous (in space) process. One can test this by using two different sizes of interval (two different N) and ascertaining that they produce the same result, or by using a process similar to the problem reconstruction technique described earlier in this chapter.

Once again the reader is strongly encouraged to learn how to program. It is a rare computing center that can be relied upon to solve your problems and using someone else's programs is usually more like wearing his shoes than courting his wife.

Problems

1. Solve the following numerically, using $\Delta t = 1$ sec. Do so by filling in the missing table entries. Carry out four significant figures.

$$\frac{dC(t)}{dt} = 0.1 \quad tC(t), \qquad C(0) = 2$$

t	C_{old}	tC_{old}	ΔC	C_{new}
0	2	0	0	2
1	2	2	0.2	2.2
2	2.2	4.4	0.44	2.64
3	—	—	—	—
4	—	—	—	—
5	4.805			

The result at 5 sec calculated in this manner is quite far from the correct value, which is 6.98. Can you explain why? Try a reverse computation on the foregoing.

2. Write and run a computer program to do Problem 1 using time steps of smaller than 1 sec. You should find that by using steps of 0.1 sec you get $C(5) = 6.675$.

3. Solve the following differential equation numerically for ten steps of t each of value 0.01. It is simple enough to do by hand to four significant figures.

$$\frac{dy}{dt} = 100 + y, \qquad y(0) = 0$$

The exact value to four figure accuracy is $y(0.1) = 10.52$. Using your result above, work backwards to time zero. You will find that at time zero you will have an error of

$$y(0) = -0.12$$

4. The following two differential equations arise in population studies. It is suggested that a digital computer be used for these two problems.
The logistic equation

$$\frac{dN}{dt} = \frac{K - N}{K} NR$$

Solve for

$$\begin{aligned} N(0) &= 10 \\ R &= 0.01 \\ K &= 300 \end{aligned}$$

for the range of t from 0 to 1000.
Try time steps of 0.1, 1, and 5.
Run with steps of 0.1, the following results were obtained

$$\begin{aligned} N(0) &= 10 \\ N(200) &= 60.9 \\ N(400) &= 195.9 \\ N(600) &= 279.9 \\ N(800) &= 297.1 \\ N(1000) &= 299.6 \end{aligned}$$

Plot your results.

5. The Van der Pole equation. Solve for the range of t from 0 to 20 in steps of 0.1 with the following values of constants

$$\frac{d^2y}{dt^2} - \lambda(1 - y^2)\frac{dy}{dt} + y = 0$$

$$\frac{dy(0)}{dt} = 0$$

$$y(0) = 0.1$$
$$\lambda = 0.4$$

6. In Chapter V we solved the two compartment problem with input beginning at time zero by making a number of simplifying approximations. Even with these approximations the solutions were still quite complex. Write a computer program for tabulating solutions in the general case.

VII

Regulation and Oscillation

1. The Sine and Cosine Functions

In elementary trigonometry the sine and cosine functions are initially defined in terms of the two angles of the right triangle that are less than 90°. The sine of the angle is the length of the side opposite the angle divided by the length of the hypotenuse, while the cosine of the angle is the length of the side adjacent to the angle divided by the length of the hypotenuse (Figure 7-1). These definitions are then extended to angles between 90° and 180° by defining the adjacent side in the negative direction from the origin, and to negative angles by defining the opposite side as being negative in the downward direction, as in Figure 7-2. They are still further extended to angles greater than plus or minus 180° by adding or subtracting multiples of the angle corresponding to a full circle, 360°. Thus, we have the following rules.

Rule: If $90° < \theta < 180°$, then

$$\sin\theta = \sin(180° - \theta)$$

$$\cos\theta = -\cos(180° - \theta)$$

Rule: For negative angles,

$$\sin -\theta = -\sin(+\theta)$$

$$\cos -\theta = \cos(+\theta)$$

There is one important difference between the way in which the sine and cosine functions are used in elementary trigonometry and the way in which they are used in calculus: in calculus, angles are almost always defined in terms of radians rather than degrees, with 2π radians being a full circle. The

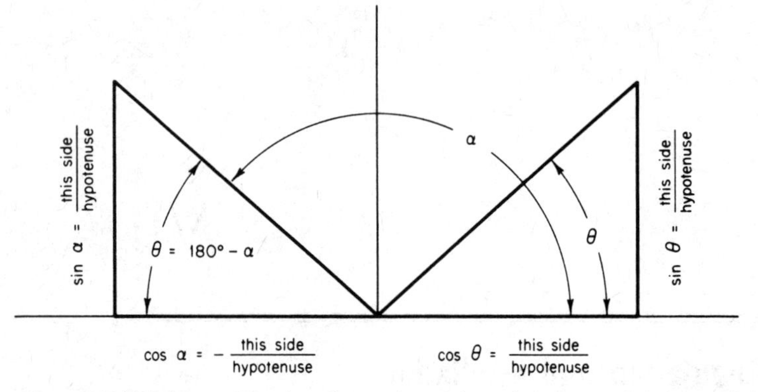

Figure 7-1. Definitions of the sine of an angle and the cosine of an angle in terms of the sides of a right triangle for angles less than and greater than 90°.

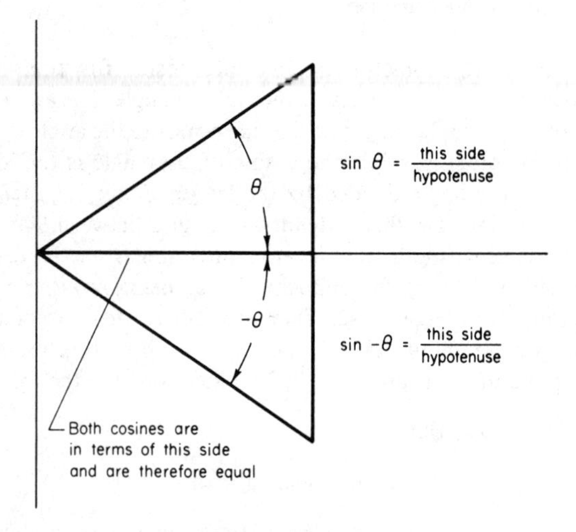

Figure 7-2. Definition of a sine of an angle and the cosine of an angle in terms of negative angles.

reason for this has to do with their derivatives. If the angles are defined in terms of radians, the derivatives of sine and cosine are given by

$$\frac{d}{d\theta} \sin \theta = \cos \theta \tag{7.1}$$

$$\frac{d}{d\theta} \cos \theta = -\sin \theta \tag{7.2}$$

and by the chain rule

$$\frac{d}{d\theta}\sin k\theta = k\cos k\theta \tag{7.3}$$

$$\frac{d}{d\theta}\cos k\theta = -k\sin k\theta \tag{7.4}$$

These derivatives are derived in Appendix III.

In Figure 7-3 we have plotted the values of the sine and cosine of an angle θ from θ equal to -3π radians to $+3\pi$ radians.

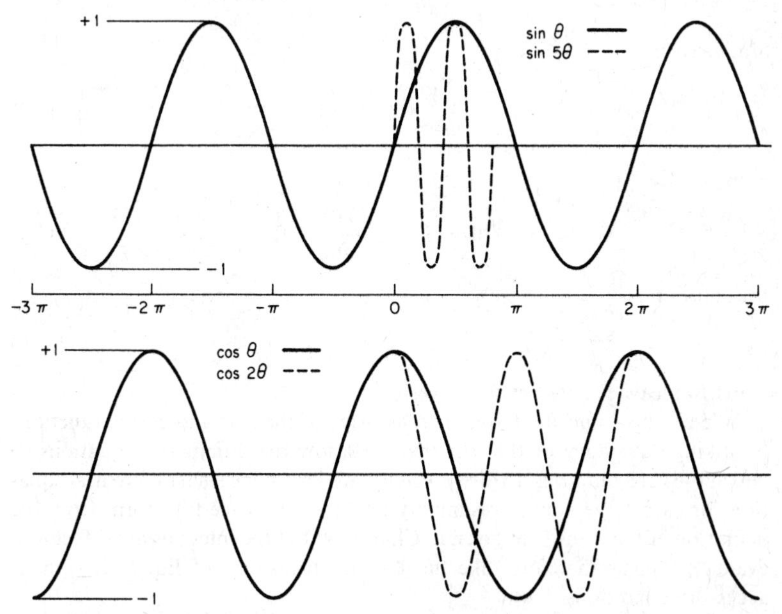

Figure 7-3. Plots of the sine and cosine functions as functions of angles shown as solid lines. The sine of five times the angle and the cosine of twice the angle are shown as dotted lines.

The importance of the sine and cosine functions in this book lies not in their geometrical interpretation but rather in the fact that they are solutions to an important class of differential equations. If

$$\frac{d^2 Y}{dt^2} + k^2 Y = 0 \tag{7.5}$$

and

$$Y(0) = 0, \qquad \frac{dY(0)}{dt} = 1 \tag{7.6}$$

the solution is

$$Y = \frac{1}{k} \sin kt \tag{7.7}$$

which we verify by using Eqs. (7.3) and (7.4)

$$\frac{dY}{dt} = \cos kt, \qquad \frac{d^2 Y}{dt^2} = -k \sin kt = -k^2 Y$$

Similarly, if

$$\frac{d^2 Y}{dt^2} + k^2 Y = 0 \tag{7.8}$$

and

$$Y(0) = 1, \qquad \frac{dY(0)}{dt} = 0 \tag{7.9}$$

the solution is

$$Y = \cos kt \tag{7.10}$$

which we leave for the reader to verify.

We can now find the Laplace transforms of the sine and cosine functions by taking advantage of the fact that we know the differential equations to which they are solutions. (More precisely, we know one such differential equation for each.) We could presumably find their Laplace transform from the definition of the transform given in Chapter 4, but the integrals are difficult to evaluate. Let us therefore take the Laplace transform of Eq. (7.5) with its accessory conditions (7.6),

$$s^2 g(s) - s\,Y(0) - \frac{dY(0)}{dt} + k^2 g(s) = 0$$

$$s^2 g(s) - 1 + k^2 g(s) = 0$$

solve for g, and since we know that the inverse of g is the sine function, we have

$$g(s) = \frac{1}{s^2 + k^2} = L\left[\frac{1}{k} \sin kt\right] \tag{7.11}$$

We leave as an exercise the derivation, using the same method, of the Laplace transform of the cosine, which is

$$g(s) = \frac{s}{s^2 + k^2} = L[\cos kt] \tag{7.12}$$

In this chapter we shall encounter a second-order differential equation whose solution for certain values of the constants is approximately a simple decaying exponential, while for other values of the constants the solution varies sinusoidally with time. It seems plausible, then, that for the intermediate values of the constants the solutions will be a compromise between these two extremes. Let us therefore examine functions of the type

$$f_1(t) = e^{-\alpha t} \sin \omega t, \qquad f_2(t) = e^{-\alpha t} \cos \omega t$$

to see what kinds of differential equations they satisfy, and by means of these establish the Laplace transform of f_1 and f_2. We begin with f_1, which we differentiate twice.

$$\frac{df_1}{dt} = e^{-\alpha t}(\omega \cos \omega t - \alpha \sin \omega t)$$

$$\frac{d^2 f_1}{dt^2} = e^{-\alpha t}[(\alpha^2 - \omega^2) \sin \omega t - 2\alpha\omega \cos \omega t]$$

Let us attempt to construct a simple differential equation whose solution will be f_1. Note that one cannot do this using only f_1 and its first derivative. The derivative contains a cosine term but f_1 does not. In the same sense in which one cannot have the sum of two different exponentials equal to zero, one cannot have the sum of a sine and cosine term equal to zero. This suggests that we try a sum of f_1 and its first and second derivatives. We see that by adding to the second derivative 2α times the first derivative, we get rid of the cosine term. The sine terms are then canceled by adding the quantity

$$(\alpha^2 + \omega^2) f_1$$

from which we find

$$\frac{d^2 f_1}{dt^2} + 2\alpha \frac{df_1}{dt} + (\omega^2 + \alpha^2) f_1 = 0$$

which is a differential equation to which f_1 is a solution. Now let us apply the Laplace transform to this, noting that

$$f_1(0) = 0, \qquad \frac{df_1(0)}{dt} = \omega$$

The transformed equation is

$$s^2 g_1 - \omega + 2\alpha s g_1 + (\omega^2 + \alpha^2) g_1 = 0$$

from which we find the Laplace transform of f_1 to be

$$g_1(s) = \frac{\omega}{s^2 + 2\alpha s + (\omega^2 + \alpha^2)} = L[e^{-\alpha t} \sin \omega t]$$

We leave it for the reader to establish that f_2 is also a solution to the same differential equation. However, in applying the Laplace transform for f_2, we must use the values

$$f_2(0) = 1, \qquad \frac{df_2(0)}{dt} = -\alpha$$

Thus

$$s^2 g_2 - s + \alpha + 2\alpha(s g_2 - 1) + (\omega^2 + \alpha^2) g_2 = 0$$

from which the Laplace transform of f_2 is found to be

$$g_2 = \frac{s + \alpha}{s^2 + 2\alpha s + (\omega^2 + \alpha^2)} = L[e^{-\alpha t} \cos \omega t]$$

Note the resemblance between these transforms and those discussed in Chapter IV, Section 10. This is the case, excluded there, in which p^2 is less than $4q$. For p^2 less than $4q$

$$L^{-1}\left[\frac{s + (p/2)}{s^2 + ps + q}\right] = e^{-(p/2)t} \cos t\sqrt{q - \left(\frac{p}{2}\right)^2} \tag{7.13}$$

$$L^{-1}\left[\frac{1}{s^2 + ps + q}\right] = e^{-(p/2)t} \frac{1}{\sqrt{q - (p/2)^2}} \sin t\sqrt{q - \left(\frac{p}{2}\right)^2} \tag{7.14}$$

2. Properties of Control Systems

In attempting to describe biological control systems, we are confronted with the fact that most of them are extremely complicated, both in their mechanism and in the description of what they do. Furthermore, many of them are interactive and no one system illustrates all of the properties of control systems we would like to study.

In order to separate the complexities of real biological systems from the principles of control theory, we illustrate a very simple control system, one that is designed to maintain a fixed level of liquid in a tank. We apologize for the fact that it is not a biological example in the strict interpretation, but point out that it is exactly analogous to the maintenance of a metabolic pool.

There are a number of characteristics of a control system that describe how well it works. If the system is disturbed in some way, how well is it able to compensate for the disturbance? Over what range is control possible? How long does it take to approximately restore the controlled variable to its proper value? Is the control system stable, or can it, in some circumstances, overcorrect, thereby producing a worse error than the one it was trying to correct? In the limiting case of this, can it get into an oscillatory condition?

We introduce some new terminology: By the *controlled variable* we mean that variable of the system which we wish to control. For example, in the case of a tank of liquid, one might choose to control the rate of flow through the tank or the level in the tank. In this case we choose to control the level in the tank. The *set point* designates that value of the controlled variable that we want to be maintained. The *unit impulse disturbance response* denotes the amount by which a controlled variable differs from the set point as a result of a unit impulse disturbance to the input of the system. The *input rate error response* is a measure of the error produced in the controlled variable by a change in input rate; the *output rate error response*, the error due to a change in output rate.

3. A Simple Regulator

Assume we have a tank of water that is being filled at a rate R_A through pipe A, and emptied at a rate R_B through pipe B. Our object is to maintain the level of liquid in the tank at a fixed level L_0 in spite of changes that might occur in R_B over which we have limited control. Clearly, the easiest way to do this would be to adjust the system in such a way that R_A is exactly equal to R_B, and to fill the tank initially to a level L_0. Since inflow and outflow rates are equal, the level in the tank does not change, and one can very easily do this by tying pipe B back to pipe A via a pump, in which case all of the liquid that flows out of the tank comes back into the tank, and on the average, neglecting cycle action of the pump, there is no change in the level. This is a system that regulates itself and, in spite of its resemblance to the cardiovascular system, is a relatively uninteresting one since, in fact, the cardiovascular system does not do this, but rather changes the total fluid volume in the system by drawing upon its various reserves in cases of stress. More realistically we must consider what happens to the level if R_B changes in relation to R_A. Clearly, if rate R_B becomes slightly less than rate R_A, the volume in the tank continually increases until the tank overflows. Conversely, if rate R_B becomes greater than rate R_A, the tank empties. Let us see what we can do about this. The first thing that comes to mind is to punch a second hole in the tank at a height L_0 above the bottom and to adjust the flow rates so that R_A is always greater than the maximum possible value of R_B. In this way, if the tank starts empty, it fills to the level L_0, at which point overflow begins and the level of the tank becomes relatively insensitive to changes in R_B. This is a variety of regulation. It is regulation by overflow and although it works, it is grossly inefficient in the sense that one must do the work to fill at a rate R_A to maintain a smaller output R_B, and in fact, regulation of this type appears to be almost nonexistent biologically. To be more precise, although overflow regulation does occur biologically, it is usually the result of some other control mechanism's being overwhelmed. For example, over the wide ranges of intake of sugar, essentially no sugar appears in the urine. However, a really massive ingestion of sugar will overwhelm the body's ability to produce insulin and overflow control then takes over.

Let us now become a little more sophisticated and try a regulating system in which the rate of input is controlled by the level in the tank. We can easily accomplish this by placing in the tank a float that operates a switch, and having the switch operate a motor that drives the pump that produces flow rate R_A. Thus, each time the level of the tank begins to fall, the motor comes on until the level is restored. If the level rises due to a decrease in R_B, the motor turns off until sufficient emptying occurs to turn the motor back on. This kind of control has, however, what appears to be a serious disadvantage to living systems, namely, that the motor is either off or on at full speed and the resulting stress on the organism is apparently biologically undesirable because such systems are biologically rather rare. The all or nothing action potential appears to be an exception to this but in fact is not. The reason it is not an exception is that the information is not usually carried by a single action potential but by the rate of firing.

Biological systems have evolved more in the direction of proportional control, which differs from the above in that the motor is not turned on or off, but rather has its speed regulated by the level in the tank. We can easily incorporate this into our model by having the float control a rheostat, which in turn controls the speed of the motor (Figure 7-4). Let us analyze this case by making the assumption that the speed of the motor is proportional to the difference between a level L_{ZF} and the actual level of water in the tank L. Thus, we may write the equation for the rate of flow R_A as

$$R_A = W(L_{ZF} - L)$$

where L_{ZF} is the level at which zero flow occurs through A, and W represents a proportionality factor between the difference between L_{ZF} and L and rate of flow R_A.

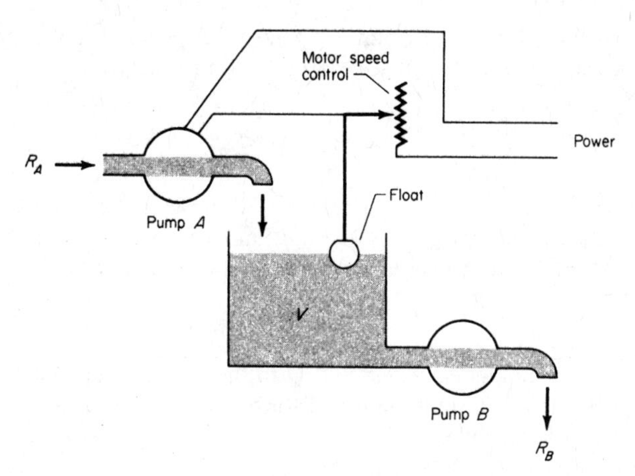

Figure 7-4. A simple regulator in which the level of liquid in a tank moves a float that controls the speed of pump A.

Now let us express the level in terms of volume. Assuming uniform cross-sectional area a of the tank, the volume at any level is given by

$$V = aL$$

so that

$$R_A = \frac{W}{a}(V_{ZF} - V) = G(V_{ZF} - V) \tag{7.15}$$

where G is W/a and has units of reciprocal seconds ($\sec^{-1}$), and V_{ZF} is the volume that causes R_A to be zero. Thus, we may write a very simple differential equation, which describes the rate of change of volume with respect to time in terms of the rate of inflow and the rate of outflow, and in turn, in terms of the rate of outflow and the volume

$$\frac{dV}{dt} = R_A - R_B = G(V_{ZF} - V) - R_B$$

or

$$\frac{dV}{dt} + GV = GV_{ZF} - R_B \tag{7.16}$$

The steady-state solution of this equation is given by

$$V_{ss} = V_{ZF} - \frac{R_B}{G} \tag{7.17}$$

Thus, to maintain V at the set point V_0 in the presence of a steady outflow R_B, we set the zero point of the float rheostat device so that

$$V_{ZF} = V_0 + \frac{R_B}{G} \tag{7.18}$$

In order to evaluate the performance of this system let us define a unit impulse disturbance analogous to the unit impulse input of Chapter V. We assume that the system has been resting in its steady state with

$$V = V_0, \qquad R_A = R_B$$

At time zero we add an extra quantity Q to the tank, perhaps by pouring a cup of water into it. The unit impulse disturbance response $H(t)$ is defined as V'/Q where V' is the difference between the actual volume and the set point volume.

Let Q be the extra volume, so that at time zero

$$V = V_0 + Q$$

and after time zero from Eqs. (7.16) and (7.18)

$$\frac{dV}{dt} + GV = GV_{ZF} - R_B = GV_0$$

These equations are somewhat simpler if written in terms of V':

$$V' = V - V_0, \qquad \frac{dV'}{dt} = \frac{dV}{dt}$$

Thus,

$$\frac{dV'}{dt} = -GV' \qquad \text{and} \qquad V'(0) = Q$$

from which we obtain

$$V'(t) = Qe^{-Gt} \tag{7.19}$$

$$H(t) = \frac{V'}{Q} = e^{-Gt} \tag{7.20}$$

Thus, the effect of adding Q is to produce a temporary change in V from which the system recovers by changing R_A with a rate of recovery determined by G (Figure 7-5).

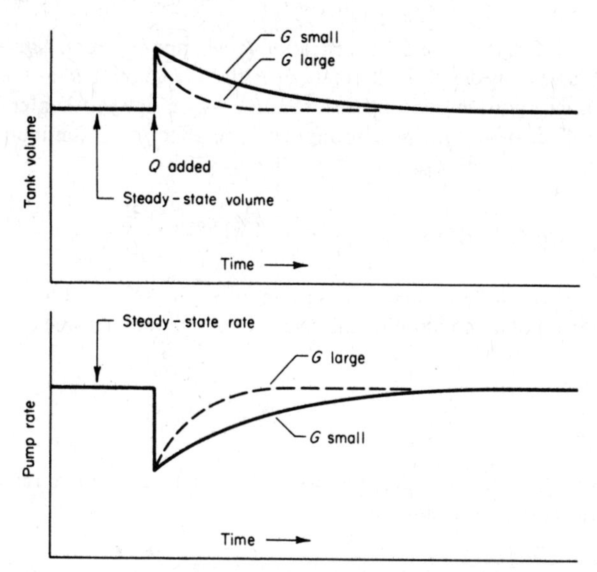

Figure 7-5. Tank volume versus time and pump rate versus time following an impulse disturbance Q.

4. The Output Rate Error Response

Suppose we had wanted to find instead the effect of decreasing R_B by an amount R'. We can do this in either of two ways. We can modify Eq. (7.16) by decreasing R_B and using the initial condition that V equals V_0, from which we obtain

$$\frac{dV}{dt} + GV = GV_{ZF} - (R_B - R') = GV_0 + R'$$

$$V(0) = V_0$$

or in terms of V'

$$\frac{dV'}{dt} = -GV' + R'$$

$$V'(0) = 0$$

so that

$$V'_{R'} = \frac{R'}{G}(1 - e^{-Gt})$$

$$V_{R'} = V_0 + \frac{R'}{G}(1 - e^{-Gt}) \tag{7.21}$$

where the subscript R' means the effect of R' as opposed to Q. Alternatively, we could have considered the decrease in R_B by the amount R' as the equivalent of a continuous extra input at a rate R'. As we saw in Chapter V, to go from the effect of a unit impulse input to the effect of a continuous input beginning at t zero, we evaluate

$$V'_{R'} = \int_0^t H(\tau)R'(t-\tau)\,d\tau = R'\int_0^t e^{-G\tau}\,d\tau = \frac{R'}{G}(1-e^{-Gt})$$

Now, let us compare the effect of a unit impulse Q to the effect of a change in R_B. In the case of a unit impulse, the original volume was restored by changing R_A by

$$R_A = R_B + \frac{dV}{dt} = R_B - GQe^{-Gt}$$

In the case of a change in R_B, the original volume is not quite restored but remains in error by the amount

$$V'_{R'}(t) = \frac{R'}{G}(1-e^{-Gt})$$

which, as t becomes large, becomes (Figure 7-6)

$$V'_{R'}(\infty) = \frac{R'}{G} \tag{7.22}$$

Note, however, that we can make this error very small by making G large and that, furthermore, by doing so we make the time required for the correction small as well. Thus, for good regulation we want G as large as possible.

The required change in R_A is given by

$$R_A = (R_B - R') + \frac{dV}{dt} = R_B + R'(e^{-Gt} - 1)$$

Thus, in either of the foregoing cases R_A is required to change by only a moderate amount rather than the full on–off swing required by our earlier model.

Let us now investigate what happens to a system of this nature if the characteristics of the control loop—the float, rheostat, and motor—change, as they might perfectly well do, as a result either of aging of physical components in our model or of changes in a biological system, which are inevitable. The steady-state volume is given by

$$V_{ss} = V_{ZF} - \frac{R_B}{G} \tag{7.23}$$

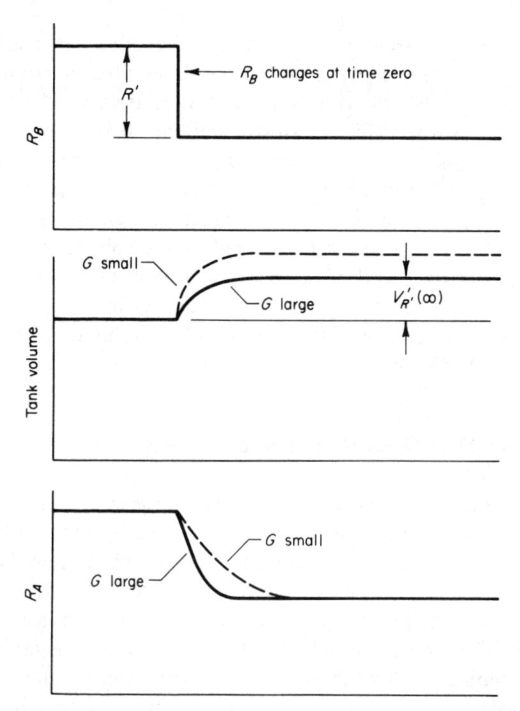

Figure 7-6. Tank volume versus time and pump rate versus time following a change in R_B.

and if our system is a good regulator, G is large and R_B/G is small compared to V_{ZF}. Thus, small changes that may occur in the ratio of R_B to G will make very little difference in the value of V. On the other hand, we must be careful about a possible change in the behavior of the sensing element, the float, which affects V_{ZF}. Suppose we place a small weight on top of the float, so that relative to the water level, the float rests at a different height. The system is designed to return the float level to the set point L_0 and has no way of knowing that the behavior of the float has changed. The system will therefore set the water level incorrectly, and this phenomenon unfortunately occurs biologically. For example, there is considerable reason to believe that blood pressure is at least partially controlled by blood flow through the kidneys, and thus diseases of the kidney that tend to increase its resistance produce a chronic increase in blood pressure. Undoubtedly, other control mechanisms, such as the carotid sinus, tend to oppose this change, but their set point appears to change over long periods of time and thus the control fails to prevent hypertension.

Except for the vulnerability of the system to change in the sensing device, it seems that by making R_B/G small we should be able to produce a system that will not only control the volume as accurately as we wish but will do so as rapidly as we want it to. This is unfortunately not true. The system we have described can correct, or correct rapidly, only for limited disturbances. If R_B exceeds GV_{ZF}, R_A cannot be increased enough to keep up with the outflow. If a very large impulse input is added, the most the control system can do is to turn off R_A and wait for the outflow to reduce the volume. Furthermore, we have assumed that the speed of the pump can be instantaneously responsive to the error in the position of the float, whereas in fact any real control system requires some time in which to respond, and this time delay limits the amount of control possible.

5. The Effect of Time Delay in the Feedback Loop

In order to analyze the effect of time delay in the system, let us investigate the simplest model of delay that we can easily build into the tank model. Let $R_A(t)$ be dependent not upon the current value $V(t)$ but upon the value of V at some earlier time $t - t_1$. We can physically realize such a delay by inserting a waterfall between the pump, which is assumed to respond instantly, and the tank (Figure 7-7). Thus, the actual rate of flow into the tank is delayed behind the pumping rate by the time required for water to fall from the pump to the tank. Assume that the system has been resting stably at its set point of volume V_0 with a flow rate $R_A(0)$; thus $V'(0)$ is zero. At time t zero, V' is changed to $+1\ \text{cm}^3$. The effect of this change will be to reduce the value of the flow from its original value $R_A(0)$ to

$$R_A = G(V_{ZF} - (V_0 + 1\ \text{cm}^3)) = R_B - G \times 1\ \text{cm}^3$$

However, this will not affect the volume in the tank until t_1 because of the delay in the waterfall. Thus, the volume in the tank remains at $V_0 + 1\ \text{cm}^3$ until t_1, when it begins to change at a rate

$$\frac{dV}{dt} = R_A - R_B = -G \times 1\ \text{cm}^3$$

Although this change will immediately affect the pump, the rate of the flow into the tank will remain at

$$R_A = R_B - G \times 1\ \text{cm}^3$$

until time $2t_1$, by which time the volume in the tank has reached the value

$$V(2t_1) = V_0 + 1\ \text{cm}^3 + (R_A - R_B)t_1 = V_0 + 1\ \text{cm}^3\ (1 - Gt_1)$$

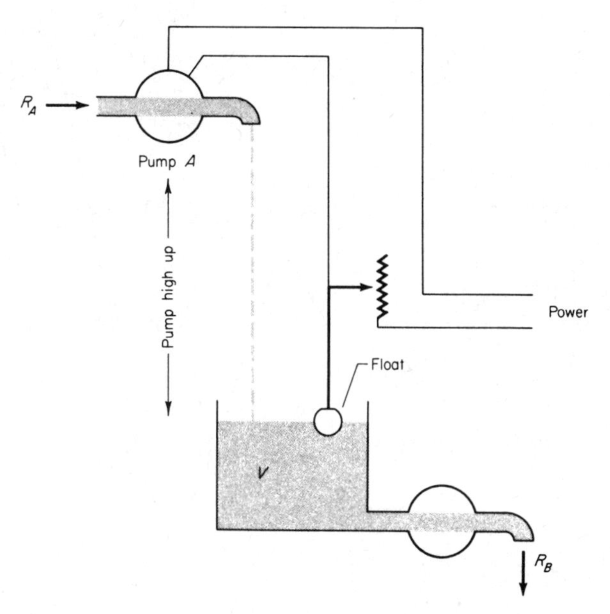

Figure 7-7. A feedback system with a waterfall delay.

If Gt_1 is less than one, the volume in the tank, at time $2t_1$ is in error by an amount less than the original 1 cm^3 and the volume is brought closer to the set point, even though it has required more time to do so than in the case in which there was no delay in the waterfall. We found, however, that in order to achieve good control, G had to be large. Suppose Gt_1 is 3. At time $2t_1$ the volume in the tank has become

$$V(2t_1) = V_0 + 1\ \text{cm}^3(1 - 3) = V_0 - 2\ \text{cm}^3$$

and we see that far from correcting the volume, the control system has produced an error larger than the original error of 1 cm^3. It has gone toward the proper correction but by the time the control system can again change the rate of flow into the tank, the volume has overshot the correct volume and has produced an error twice as big in the other direction. See Figure 7-8.

The next attempt to control the volume will then overshoot by about 4 cm^3 in the original direction and the control gets progressively worse. A moment's thought will convince the reader that the particular point at which the delay occurred in this model is irrelevant. The same difficulty would have occurred if the delay had been between the sensing device and the motor, as it might have been, for example, if they were connected by a neuron. Thus, we find

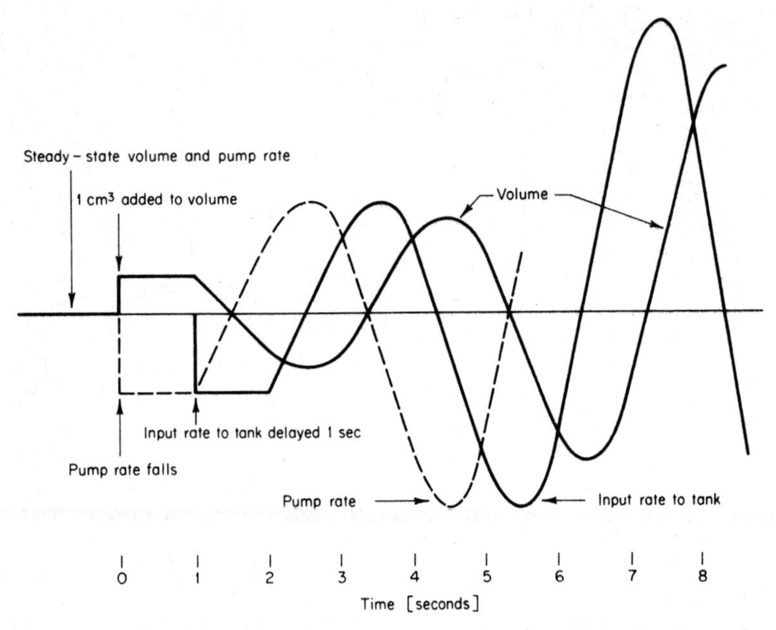

Figure 7-8. The effect of a waterfall delay of 1 sec and a value of G equal to 2.1 sec^{-1}.

that the maximum allowable value of G in the presence of delay t_1 is approximately

$$G = \frac{2}{t_1} \tag{7.24}$$

It is perhaps not obvious why Gt_1 cannot be any value less than 2. If it were 1.5, for example, the system would overshoot but would, at time $2t_1$, be in error by 0.5 cm^3 and in the opposite direction. The reason for this is that an overshoot in one direction is followed by an overshoot in the other direction. If a second disturbance, say, another cup of water, enters the system timed to coincide with the first overshoot in the direction of the original error, the effects of the two will add and the next swing will be still wider. If the system is to be able to correct arbitrary kinds of disturbances, t_1G must be less than one. Thus, the existence of delay in the system limits the amount of control possible.

The foregoing analysis is approximate, but adequate to demonstrate the principle. An exact analysis indicates that the system goes into sustained oscillation for G greater than $\pi/2t_1$.

6. Pooling Delay versus Pure Time Delay

The reader may wonder about the use of a waterfall in the previous example and whether the result was dependent upon the use of a somewhat contrived type of delay. A waterfall was chosen because it is the simplest example of a pure time delay. A pure time delay is one in which the output has exactly the same time dependence as the input but is delayed behind it. Note that a pipe would not produce such a delay because water is essentially incompressible, which implies that the rate of outflow is instantaneously the same as the rate of inflow. Indeed, devices that produce pure time delay are rare, although a neuron, in a sense, approximates this, as does the hatching of an egg in population study. More commonly, however, delay in physical and biological systems is due to pooling, and we will see that this differs in an important way from pure time delay. The simplest example of pooling delay is a tank whose outflow is proportional to the height of the liquid in the tank. A simple hole in the bottom of a tank does not have this property but a narrow pipe at the output, which limits the flow by viscous resistance, comes very close. If the input to the tank is at a rate of R_{in}, the rate of change of volume is given by

$$\frac{dV}{dt} = R_{\text{in}} - R_{\text{out}} = R_{\text{in}} - KV$$

where K is a proportionality constant.

Thus, if the input changes from R_{in_1} to R_{in_2} after being in a steady state, the output changes according to

$$R_{\text{out}}(t) = R_{\text{in}_2} - (R_{\text{in}_2} - R_{\text{in}_1})e^{-Kt}$$

which is the familiar exponential behavior.

Now consider the effect of an input that changes from R_{in_1} to R_{in_2} and back to R_{in_1}, as shown in Figure 7-9. The result resembles a time delay but the shape of R as a function of time has been changed. This kind of delay is conceptually more difficult than pure time delay but often mathematically much simpler. More importantly, it is the kind of delay that occurs most often in physical and biological systems. We have chosen to call it *pooling delay* to emphasize that it is exactly the phenomenon that occurs in metabolic pooling. It is also in many cases descriptive of inertial effects in mechanical systems. Thus, in our previous example, had we chosen to use as the delay the time it takes for a real mechanical pump to come up to speed when its input voltage is changed, pooling delay would have been a reasonably good model. To see this, imagine a phonograph that has just been turned on. The speed of the turntable does not instantly come to its final value, but approaches it approximately exponentially.

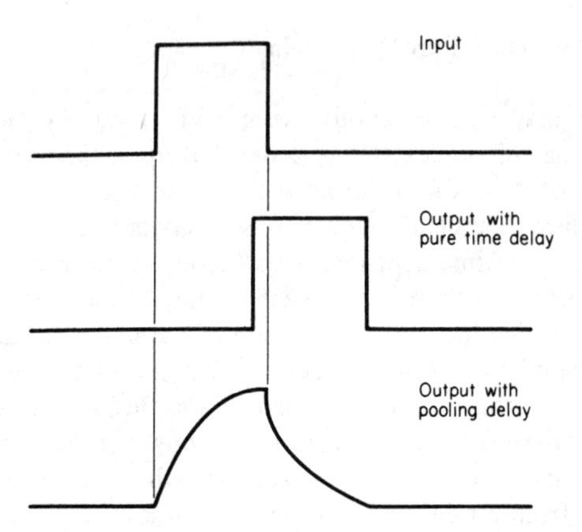

Figure 7-9. Comparison of pure time delay to pooling delay.

Now let us consider the effect of pooling delays on a feedback control system. In the first example we could have incorporated a pooling delay by considering the time required for a motor to come up to speed. Alternatively, we could have incorporated a pooling delay in the feedback "loop" by having the output of the first tank fill a second tank and monitoring the level in the second tank. Mathematically, the two situations are identical. The second case, however, is conceptually simpler, so we shall use it as a model.

Consider two tanks in series (Figure 7-10). Let the outflow of the first tank be $R_1 = K_1 V_1$ and the outflow of the second tank be R_B. The inflow to the second tank will be the outflow of the first. The inflow to the first will be controlled by the level in the second tank via the pump. Thus,

$$\frac{dV_1}{dt} = R_A - R_1 = \frac{W}{a_2}(V_{ZF} - V_2) - K_1 V_1$$

$$= G(V_{ZF} - V_2) - K_1 V_1 \tag{7.25}$$

$$\frac{dV_2}{dt} = K_1 V_1 - R_B \tag{7.26}$$

where G now equals W/a_2.

Now let us eliminate V_1 by solving (7.26) for V_1

$$V_1 = \frac{1}{K_1}\left(\frac{dV_2}{dt} + R_B\right)$$

$$\frac{dV_1}{dt} = \frac{1}{K_1}\left(\frac{d^2 V_2}{dt^2}\right)$$

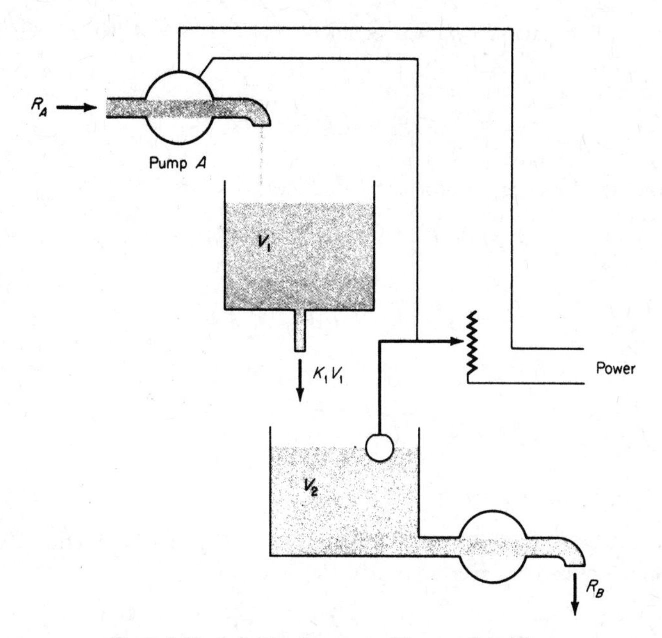

Figure 7-10. A feedback system with a pooling delay.

and insert these into Eq. (7.25)

$$\frac{d^2V_2}{dt^2} + K_1\frac{dV_2}{dt} + K_1GV_2 = K_1(GV_{ZF} - R_B) \tag{7.27}$$

Thus, the set point steady-state value is given by

$$V_{2,0} = V_{ZF} - \frac{R_B}{G} \tag{7.28}$$

In order to see the effect of a disturbance in the level V_2, let us again make use of V', the difference between the set point level and the actual level

$$V' = V_2 - V_{2,0}, \qquad \frac{dV'}{dt} = \frac{dV_2}{dt}, \qquad \frac{d^2V'}{dt^2} = \frac{d^2V_2}{dt^2}$$

Thus, from Eqs. (7.27) and (7.28)

$$\frac{d^2V'}{dt^2} + K_1\frac{dV'}{dt} + K_1GV' = 0 \tag{7.29}$$

and $V'(0) = 0$. At time zero, Q is added to V_1. Prior to this, since the system was in a steady state, dV_2/dt was equal to $K_1V_1 - R_B = 0$. Thus when Q is added to V_1,

$$\frac{dV_2}{dt} = K_1 Q = \frac{dV'}{dt}$$

Applying the Laplace transform, we find

$$s^2 g(s) - K_1 Q + K_1 s g(s) + K_1 G g(s) = 0$$

$$g(s) = \frac{K_1 Q}{s^2 + K_1 s + K_1 G} = \frac{K_1 Q}{s^2 + ps + q} \tag{7.30}$$

where

$$p = K_1 \tag{7.31}$$

and

$$q = K_1 G \tag{7.32}$$

If G is small, this form satisfies the conditions of Chapter IV, Section 10, and the transform can be inverted to find

$$V'(t) = K_1 Q \frac{e^{-bt} - e^{-at}}{a - b} \approx K_1 Q \frac{e^{-Gt} - e^{-K_1 t}}{K_1 - G} \tag{7.33}$$

$$a = p = K_1, \qquad b = q/p = G$$

which is similar to the concentration in the second compartment of a two-compartment series dilution problem. V' starts at zero, increases to a peak, and returns to zero. If G is small, however, the regulation against the effect of a change in R_B cannot be very good. To see this we calculate the unit impulse response

$$H(t) = (e^{-Gt} - e^{-K_1 t}) \frac{K_1}{K_1 - G} \tag{7.34}$$

and from this the result of changing R_B by the amount R'

$$V'_{R'} = \int_0^t H(\tau) R'(t - \tau)\, d\tau = \frac{R'K_1}{K_1 - G} \left(\frac{1 - e^{-Gt}}{G} - \frac{1 - e^{-K_1 t}}{K_1} \right)$$

which, as t becomes large, becomes

$$V'_{R'}(\infty) = \frac{R'}{G} \tag{7.35}$$

so that for small G we have poor regulation.

To find the inverse of $g(s)$ for large values of G we cannot use the forms given in Chapter IV since these required that p^2 be greater than $4q$, which would not be the case for large values of G. To find a clue to the inversion of this case, let us consider the special subcase in which G is large but K_1 small, so that we can neglect the quantity sK_1 in the denominator of g. When we do so, we find that g is in one of the forms derived in Section 1, Eq. (7.11).

$$g(s) = \frac{K_1 Q}{s^2 + K_1 G}$$

Thus, its inverse is

$$V' = \frac{K_1 Q}{k} \sin kt \tag{7.36}$$

where

$$k = \sqrt{K_1 G}$$

The effect of adding Q to V_1 in this special subcase is to cause V' to oscillate sinusoidally around zero with an amplitude

$$\text{amp} = \frac{K_1 Q}{k} = Q\sqrt{\frac{K_1}{G}}$$

Using this clue, we return to the case where K_1 is not necessarily negligible, but G is large. Since for small G we found a double exponential-type solution, Eq. (7.33), and for large G a sinusoidal solution (7.36), it seems reasonable to guess that for intermediate values of G the solution might be a compromise between these. It suggests that we try a function that starts at zero, rises to a peak, comes back toward zero, passes through zero, and perhaps oscillates in a decaying manner thereafter. Let us therefore try

$$e^{-\alpha t} \sin \omega t$$

whose Laplace transform we derived in the introduction.

Referring to Eq. (7.14) we find that

$$V' = \frac{K_1 Q}{\sqrt{q - (p/2)^2}} e^{-(p/2)t} \sin t\sqrt{q - \left(\frac{p}{2}\right)^2}$$

is the inverse of $g(s)$ in the case in which p^2 is less than $4q$. Now for G sufficiently large

$$\sqrt{q - (p/2)^2} = \sqrt{(K_1 G) - \tfrac{1}{4}K_1{}^2} = \sqrt{K_1 G}$$

Thus, V' is approximately

$$V' \approx Q\sqrt{\frac{K_1}{G}}\, e^{-K_1 t/2} \sin t\sqrt{K_1 G} \tag{7.37}$$

This function oscillates around zero with a decaying amplitude. Its peak value, which occurs on the first cycle at time t given approximately by

$$t\sqrt{K_1 G} = \frac{\pi}{2}$$

is

$$V'_{\text{peak}} = Q\sqrt{\frac{K_1}{G}}\, e^{-(\pi/4)\sqrt{K_1/G}}$$

Thus, we find that the response to the addition of Q to V_1 is

$$V' = Q\sqrt{\frac{K_1}{G}}\, e^{-K_1 t/2} \sin t\sqrt{K_1 G} \tag{7.38}$$

which we have plotted in Figure 7-11. The phenomenon shown here is called

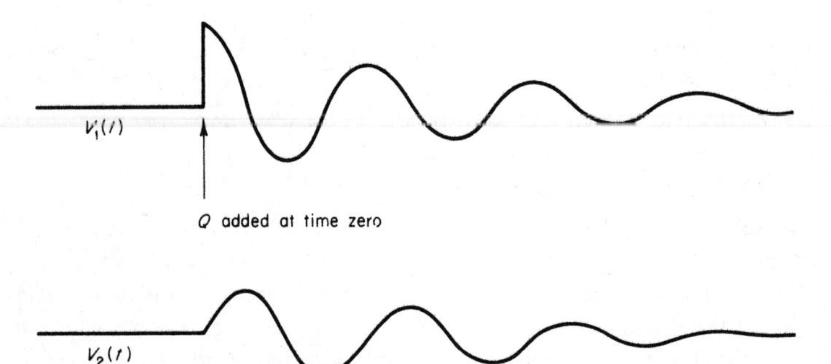

Figure 7-11. Volumes in the two tanks plotted as functions of time following an impulse disturbance Q in the first tank.

overshoot. In trying to correct the volume, the control system has overshot the proper setting because of the pooling delay. It is a commonly occurring phenomenon in both physical and biological control systems but is not necessarily objectionable since the controlled variable finally settles down to the correct value. It might be objectionable if the disturbance occurred periodically, as described in the earlier example of the effect of time delay in the feedback loop, or if the organism were particularly sensitive to its effects. The latter case is observable in the hypoglycemia that occasionally follows a large ingestion of sugar (12). (In this case, the two compartments are represented by storage of insulin in the pancreas and storage in the blood.)

The maximum value G can have for zero overshoot is that for which p^2 equals $4q$. From (7.31) and (7.32) we find this will occur when

$$G_{\text{m}} = \frac{K_1}{4} \tag{7.39}$$

In this case Eqs. (7.30) reduce to

$$g(s) = \frac{K_1 Q}{(s + (p/2))^2} \tag{7.40}$$

the inverse of which is

$$V'(t) = QK_1 te^{-(p/2)t} = QK_1 te^{-(K_1/2)t} \tag{7.41}$$

Thus, the unit impulse disturbance response is

$$H(t) = K_1 te^{-(K_1/2)t} \tag{7.42}$$

from which we find the effect of reducing R_B by R' to be

$$V'_{R'} = \int_0^t K_1 \tau e^{-K_1\tau/2} R'(t - \tau) d\tau$$

which as $t \to \infty$ becomes

$$V'_{R'}(\infty) = \frac{4R'}{K_1} = \frac{R'}{G_m} \tag{7.43}$$

We can, however, have considerably better control over the system if we allow some overshoot. Let us calculate the effect of a change in R_B by the amount R' for large values of G. From Eq. (7.38) we can find the unit error impulse response by dividing by Q.

$$H(t) = \sqrt{\frac{K_1}{G}}\, e^{-K_1 t/2} \sin t\sqrt{K_1 G}$$

From this we can find the effect of a change R', which is

$$V'_{R'} = \int_0^t H(\tau)R'(t - \tau)\, d\tau \xrightarrow[t \to \infty]{} \frac{R'}{G} \tag{7.44}$$

In Figure 7-12 we show the effect of a change R' equal to 10 cm^3/sec for G equal to 10 times G_m, the maximum value of G for zero overshoot. Compare this to the effect of a change R' equal to 2 cm^3/sec with G equal to G_m shown in Figure 7-13. Not only is the control better in the case of large G, but except for the initial overshoot, the controlled volume comes within a small error of its final value in a much shorter time than in the case of zero overshoot. Thus, we see that the effect of a pooling delay determined by K_1 is somewhat analogous to the effect of a time delay in that it limits the extent to which we can control the system if we demand zero overshoot.

In the cases where the limitation caused by overshoot is objectionable, it turns out that there is something that can be done about it. If the float can measure not only the level of liquid in the tank but its rate of change as well,

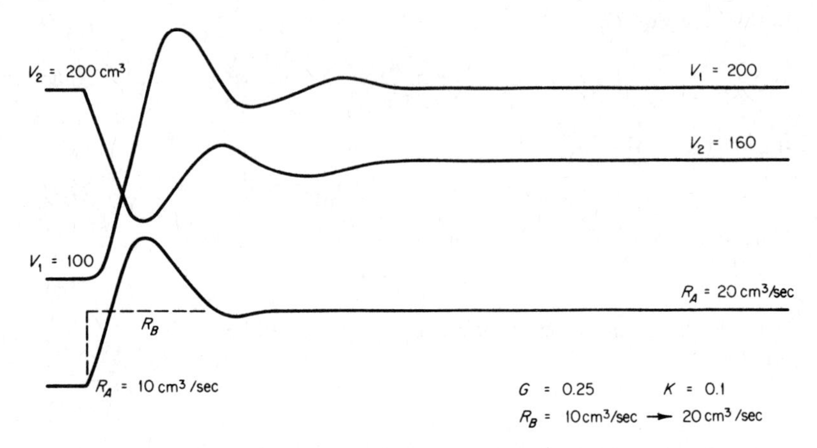

Figure 7-12. Volumes in the two tanks and pumping rate plotted against time for G equal to 10 times the zero overshoot value, and R' equal to 10 cm³/sec.

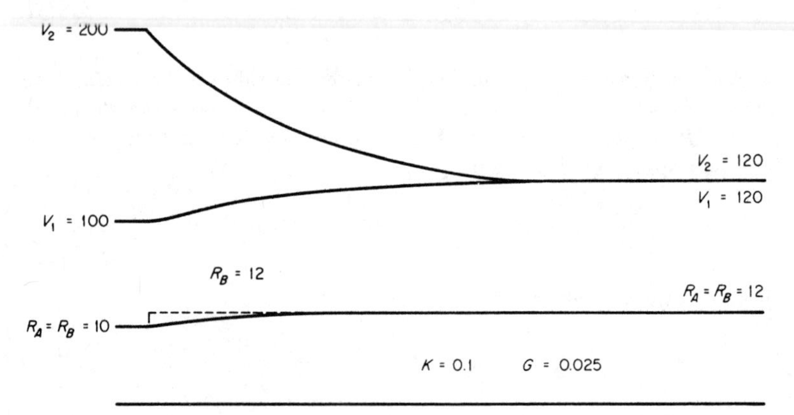

Figure 7-13. Volumes in the two tanks and pumping rate for G equal to the maximum value for zero overshoot and R' equal to 2 cm³/sec.

it is possible to modify the control system to eliminate the effects of the pooling delay. Consider Eq. (7.25). Let us modify R_A so that

$$R_A = G(V_{ZF} - V_2) - M\frac{dV_2}{dt} \tag{7.45}$$

then Eq. (7.27) becomes

$$\frac{d^2V_2}{dt^2} + K_1(1 + M)\frac{dV_2}{dt} + K_1GV_2 = (GV_{ZF} - R_B)K_1$$

and (7.29) becomes

$$\frac{d^2V'}{dt^2} + K_1(1+M)\frac{dV'}{dt} + K_1GV' = 0 \tag{7.46}$$

To solve this we could proceed in a manner analogous to the previous solution, noting that if we choose M large enough, we can avoid the situation in which $4q$ becomes greater than p^2. In a more direct manner we can see if K_1G and $K_1(1 + M)$ are both large, the effect of the second derivative term of (7.46) is negligible and the solution reduces to that of the case without pooling delay, Eqs. (7.19) and (7.21), except that G is divided by $1 + M$.

$$V' = Qe^{-[G/(1+M)]t}$$

$$V'_{R'} = \frac{R'(1+M)}{G}\left(1 - e^{-[G/(1+M)]t}\right) \tag{7.47}$$

This technique of incorporating rate of change measurement into the control of R_A is called *velocity feedback*. Its name comes from the fact that it is dependent upon the velocity of V_2 or its rate of change with respect to time. By analogy, the feedback dependent only on the value of V_2 is called *position feedback*.

Although velocity feedback is helpful in eliminating the limitation on G in a pooling delay system, it is not without difficulty of its own. The principal difficulty is that measurement of velocity is difficult. For example, small ripples on the surface of the liquid would not significantly interfere with position measurements but could be interpreted as velocity changes, which would then be magnified by the feedback system. Without going into the details, let us note that errors of velocity measurement set a limit on how effective velocity feedback can be in eliminating limitations imposed by the pooling delay.

7. A Three-Compartment System

As the reader might have anticipated, the addition of a third compartment with one more pooling delay is likely to produce a system that can accentuate the overshoot phenomenon to the point where the system can go into a self-sustaining oscillation. This condition might occur if, for some input, the effect of the controller is to replicate that input or replicate it with amplitude larger than the original.

In order to study this case let us separate the effect of the feedback from the main path by having the level of the third compartment control a pump but not use the output of that pump as the input to the system (Figure 7-14).

Instead let the input to the three compartments be a flow rate that has a component sinusoidal in time. Let

$$R_A = R_{A,0} + R_{A,1} \sin \omega t$$

$$R_B = R_{A,0}$$

where

$$R_{A,1} < R_{A,0}$$

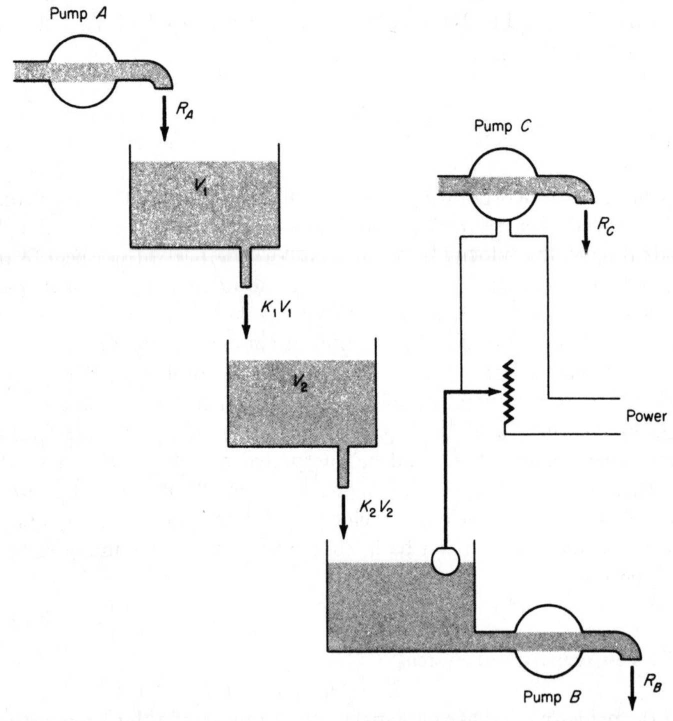

Figure 7-14. A three-compartment system in which the level in the third compartment controls the speed of pump C.

We write the equations for the three volumes based on the first two tanks losing liquid at rates proportional to their volume and the third being emptied at a fixed rate.

$$\frac{dV_1}{dt} = R_A - K_1 V_1$$

$$\frac{dV_2}{dt} = K_1 V_1 - K_2 V_2$$

$$\frac{dV_3}{dt} = K_2 V_2 - R_B$$

These are combined by solving the third one for V_2, differentiating it, and inserting the results into the second of foregoing equations.

$$V_2 = \frac{1}{K_2}\left(\frac{dV_3}{dt} + R_B\right)$$

$$\frac{dV_2}{dt} = \frac{1}{K_2}\frac{d^2V_3}{dt^2} = K_1 V_1 - K_2 V_2 = K_1 V_1 - \left(R_B + \frac{dV_3}{dt}\right)$$

This is now solved for V_1, which is inserted into the first of the three equations above.

$$V_1 = \frac{1}{K_1 K_2}\frac{d^2V_3}{dt^2} + \frac{1}{K_1}\left(R_B + \frac{dV_3}{dt}\right)$$

$$\frac{dV_1}{dt} = \frac{1}{K_1 K_2}\frac{d^3V_3}{dt^3} + \frac{1}{K_1}\frac{d^2V_3}{dt^2} = R_A - K_1 V_1$$

Thus collecting terms, we find

$$\frac{d^3V_3}{dt^3} + (K_1 + K_2)\frac{d^2V_3}{dt^2} + K_1 K_2 \frac{dV_3}{dt} = K_1 K_2 (R_A - R_B) \tag{7.48}$$

The effect of V_3 on the controller pump will be to produce a flow R_C

$$R_C = G(V_{ZF} - V_3)$$

Now let us apply an input

$$R_A = R_B + R_{A,1} \sin \omega t \tag{7.49}$$

and guess that the condition we are looking for is one that will cause the controller to produce a sinusoidal output, which would be one in which

$$V_3 = V_{3,0} + A \sin \omega t$$

and therefore

$$R_C = G(V_{ZF} - V_3) = G(V_{ZF} - V_{3,0}) - GA \sin \omega t$$

Let us try this form for V_3 by differentiating it three times,

$$\frac{dV_3}{dt} = A\omega \cos \omega t$$

$$\frac{d^2 V_3}{dt^2} = -A\omega^2 \sin \omega t$$

$$\frac{d^3 V_3}{dt^3} = -A\omega^3 \cos \omega t$$

and inserting the derivatives into Eq. (7.48)

$$(-A\omega^3 + K_1 K_2 A\omega) \cos \omega t - A\omega^2 (K_1 + K_2) \sin \omega t = K_1 K_2 (R_A - R_B)$$

and with Eq. (7.49), the expression is equal to

$$K_1 K_2 R_{A,1} \sin \omega t$$

from which we find that there exists a frequency

$$\omega^2 = K_1 K_2$$

for which

$$A = -\frac{R_{A,1} K_1 K_2}{\omega^2 (K_1 + K_2)} = -\frac{R_{A,1}}{K_1 + K_2}$$

and the output of the pump is related to the input by

$$R_C = G(V_{ZF} - V_3) = G(V_{ZF} - V_{3,0}) + \frac{GR_{A,1}}{K_1 + K_2} \sin \omega t$$

and

$$R_A = R_B + R_{A,1} \sin \omega t$$

Thus, the output of the pump consists of a steady component $G(V_{ZF} - V_{3,0})$ and a sinusoidal component that replicates the sinusoidal part of the input except that it is multiplied by the factor

$$\frac{G}{K_1 + K_2}$$

If the output of the pump is tied back to the input of the system and the factor above is greater than one, the system, once started at this frequency, will continue to oscillate. In fact, one does not usually even have to start it. Random fluctuations that always exist in any physical system will start the oscillations. Thus, a system such as described with $G/(K_1 + K_2)$ greater than one is inherently unstable.

8. Control of Biosynthesis

Earlier in this chapter we stated that the tank models were similar to the pooling of chemical compounds. To see how this comes about, let us consider the control of rates of chemical, and particularly biological chemical, reactions.

In general, the rate of a chemical reaction is dependent upon the product of the concentrations of the reacting substances with a rate constant K characteristic of the reaction. Thus, if one molecule of A combines with one of B, the rate of production of the product P per unit volume is given by

$$\frac{dP}{dt} = K|A|\,|B|$$

where the vertical bars denote concentrations.

If two molecules of A combine with one of B, the rate is given by

$$\frac{dP}{dt} = K|A|\,|A|\,|B| = K|A|^2\,|B|$$

In the simplest cases, in which a single molecule of A decomposes into two or more products, the rate of formulation of each product is

$$\frac{dP}{dt} = K|A|$$

In each of the foregoing the rate of formation of the product increases in proportion to the first or a higher power of the concentrations of the reacting substances.

When the reaction involves an enzyme, a different law appears to govern the reaction rate. Such reactions are believed to involve at least two steps: the first, in which the substrate combines with the enzyme to form an intermediate product, is usually a relatively fast reaction; it is followed by a second step, at a relatively slower rate, in which the intermediate breaks down into a final product and the enzyme molecule is released to react again with another molecule of the substrate. Although each step follows the foregoing kinetic laws, the fact that some of the enzyme is tied up in the intermediate product causes the reaction rate to be governed by the amount of free enzyme present, which is the difference between the amount initially present and the amount tied up in the intermediate product. Such reactions are usually described by the diagram

$$S + E_F \underset{k_2}{\overset{k_1}{\rightleftharpoons}} I \xrightarrow{k_3} P + E_F$$

where E_F is the free enzyme.

The first step of this reaction is indicated as being reversible. This is to be consistent with the vast body of literature in this field. In fact, it is irrelevant to most of the following discussion; furthermore, there are many cases in which the second step of the reaction is also at least partially reversible, but the foregoing is the most common description of the process.

If E_T represents the total amount of enzyme (in terms of concentrations), the amount initially put into the test tube, the concentration of free enzyme, is given by

$$E_F = E_T - I$$

Now let us apply the simple kinetic laws to each step of the reaction:

$$\frac{d|I|}{dt} = k_1 S E_F - (k_2 + k_3) I$$

$$\frac{dP}{dt} = k_3 I$$

Replacing E_F by $E_T - I$, we find in the steady state for which dI/dt is zero, that

$$I = \frac{k_1 S E_T}{k_1 S + k_2 + k_3} = \frac{S E_T}{S + (k_2 + k_3)/k_1} = \frac{S E_T}{S + K_M}$$

and the rate of generation of the product is

$$\frac{dP}{dt} = k_3 \frac{S E_T}{S + K_M}$$

where K_M is the Michaelis constant of the reaction. This is called the *Michaelis–Menten equation.* At low concentrations of the substrate the rate is approximately

$$\frac{dP}{dt} = \frac{k_3 S E_T}{K_M}$$

and this case is referred to as being substrate limited. At high concentrations of the substrate, the rate is limited by the enzyme available:

$$\frac{dP}{dt} = k_3 E_T$$

That these rates are obtained experimentally is one of the key pieces of evidence in support of the two-step model of an enzymatic reaction.

From the point of view of a control system the importance of this scheme is the following. Suppose a cell normally produces P at a rate of R_P. In order to do so, it must generate S at the same rate or faster. Now, suppose that for some reason the cell needs to turn off the production of P. It can do so in two ways. It can produce a substance called a *competitor* that reacts with the substrate and continually pulls it out of the S to P reaction. To do so, however, it must produce the competitor at the rate comparable to R_P. A far more economical thing for it to do is to produce a competitor that reacts with the enzyme, which is normally present in much smaller quantities and is generated at a much lower rate. As described above, the rate of normal generation of the enzyme can approach zero, since it is used but not destroyed. Thus, a very small amount of the competitor can control the rate of the reaction. Frequently, the competitor is a metabolic product farther down a chain of reactions from P. This is called *product inhibition*. It occurs very commonly, and its control is so effective that if a cell is placed in a bath from which a metabolic substance in the middle of a chain can diffuse into the cell, the cell will often turn off its own production of the substance farther up the chain.

Let us see how product inhibition is useful to a living organism. Consider the following chain of reactions.

$$A \xrightarrow[K_A]{E_A} B \xrightarrow[K_B]{E_B} C \xrightarrow[K_C]{E_C} D$$

To be specific, let us assume that the first step is enzyme limited but that subsequent steps are substrate limited.

D is produced at a rate (approximately)	$K_C CE_C$
C is produced at a rate (approximately)	$K_B BE_B$
B is produced at a rate (approximately)	$K_A E_A$

In the steady state

$$\frac{dD}{dt} = K_C CE_C = K_B BE_B = K_A E_A$$

Now suppose for some reason the cell needs to turn off the production of D. It can do so by reducing the E_C, but if nothing else changes, the result of this is to cause the concentration of C to increase, which is not only wasteful of the cell's energy but makes it difficult to reduce the production of D. A reduction in E_C is accompanied by an increase in C, so that their product remains constant, a fact that is evident from the foregoing equation. Now suppose that C reacts reversibly with E_A. The amount of E_A available for the A to B step is reduced approximately in proportion to the amount of C present. This is not exact, since the reaction is reversible, but is close enough for small concentrations of C to establish the principle. Let

$$E_A = E_{A,0} - E_{A,1}C$$

where $E_{A,0}$ is the amount of E_A available if C is zero and $E_{A,1}$ is a proportionality constant. Now

D is produced at a rate	$K_C CE_C$
C is produced at a rate	$K_B BE_B$
B is produced at a rate	$K_A(E_{A,0} - E_{A,1}C)$

In the steady state

$$K_A(E_{A,0} - E_{A,1}C) = K_C CE_C$$

$$C = \frac{K_A E_{A,0}}{K_A E_{A,1} + K_C E_C}$$

$$\frac{dD}{dt} = \frac{K_A E_{A,0} K_C E_C}{K_A E_{A,1} + K_C E_C}$$

If $K_A E_{A,1}$ is large compared to $K_C E_C$, the cell can reduce the production of D by reducing E_C.

9. Other Oscillators

We have seen how a feedback control system can oscillate, but not all oscillators involve feedback control. Most biological oscillators are of a much simpler type, called *relaxation oscillators.* The simplest example of a relaxation oscillator is a dripping faucet, and the most obvious biological example is a pacemaker cell.

Both nerve and muscle cells maintain their inside electrical potential about 70 mV negative with respect to the outside environment. They do so by actively pumping sodium out of the cell and potassium into the cell. A steady state exists when the rate of passive diffusion in the opposite directions equals the pumping rate. If for some reason the potential difference decreases it causes the cell membrane to become more permeable to sodium and the steady state is upset by an influx of sodium ions, which in turn further reduces the membrane potential difference. This causes a still further increase in permeability and a further decrease in membrane potential difference, until the inside of the cell comes to within a few millivolts of the outside. This is the mechanism of an action potential. The effect of the potential change on the permeability lasts only a few milliseconds, after which the permeability decreases, causing the cell to return to its normal 70-mV difference, and the steady state is reestablished until something triggers the cell again.

In most nerve and muscle cells the trigger is some external action. In the case of the pacemaker cells, after returning to the steady state the cell begins to leak sodium, so that it slowly drifts toward a threshold voltage, at which point the membrane permeability starts changing and the cycle repeats.

Like the dripping faucet, pacemaker action depends upon some parameter of the system reaching a threshold value, at which point some property of the system changes. In the case of the dripping faucet the threshold is the weight of the drop exceeding the surface tension force. In the case of the pacemaker the threshold is the membrane potential at which the permeability starts changing.

There are still other very simple systems that will oscillate. Consider the following pairs of reactions in which A and B are formed and destroyed at rates that are independent of their own concentrations.

$$\xrightarrow{R} A \xrightarrow{R_A'}$$

$$\xrightarrow{R_B} B \xrightarrow{R_B'}$$

This independence is not essential to the argument but makes it simpler. Let the rate constant of formation of A be inhibited by the pool of B. Furthermore, let the rate constant of destruction of B be inhibited by the pool of A:

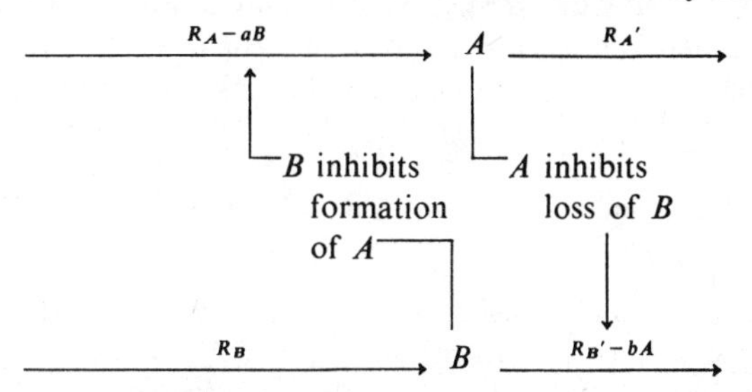

This system can potentially oscillate. It does not have to oscillate, depending upon the initial conditions, but it can do so and we have modeled a case that does numerically in Section 4 of the last chapter.

Still another simple oscillating system is the ring oscillator. Imagine two neurons connected head to tail. Each time one fires it will trigger the other at a slightly later time. Thus, the system once started continues to oscillate. This kind of oscillation has been suggested as the mechanism of short-term memory. While there does not seem to be any definite evidence for this, it has a number of properties that one would require of a short-term memory system. It is easily started: A single incoming action potential can start the loop oscillating. It requires no chemical change in the cells other than those that normally occur on cell firings. In the event of traumatic injury to the system one would expect that the oscillations might be interrupted and, of course, they would not restart themselves, which would result in loss of memory of events immediately preceding the trauma. This is often observed clinically.

For detailed descriptions of biological oscillators, particularly their relationship to circadian rhythms, see (1, 3, 4, 13a, 14, 16, 21, 26, 30). For further discussion of control systems in general, see (1, 12, 13, 16, 19, 20, 22, 28, 29, 31).

Problems

1. Given a tank of 100 cm^2 cross-sectional area with $R_B = 10\ cm^3/sec$, find V_{ZF} for $G = 3\ sec^{-1}$ when the set point volume is 1000 cm^3. What is V if $R_B = 0$?

2. In Problem 1 change R_B from 10 to 11 cm^3/sec. How much does the volume change? How much would it change if $G = 1\ sec^{-1}$?

3. Using the tank in Problem 1, assume $R_B = KV$, $K = 0.1\ sec^{-1}$. Write the equation that describes this system. What is V_{ZF}? What happens if Q is added at time zero?

4. In Sections 3 and 4 we used as a model a tank whose inflow was feedback controlled and whose output varied. We could equally well have controlled the output pump by the level in the tank and varied R_A. Assume

$$R_B = G(V - V_{ZF})$$

If $G = 3\ sec^{-1}$ and the set point is $V = 1000\ cm^3$, what is V_{ZF} for $R_A = 10\ cm^3/sec$?

5. Consider a two-tank model such as is discussed in Section 6. Instead of using a pump to control R_B, assume $R_B = K_2 V_2$. Write the equation that describes this system. Show that the condition for zero overshoot is

$$G \leq \frac{(K_1 - K_2)^2}{4K_1}$$

VIII

Diffusion

1. Introduction: Infinite Series

The solutions of some of the problems in this chapter are described in terms of infinite series. An infinite series consists of a sum of an infinite number of terms. Thus, for example, the number e can be expressed as the sum of the following series, where the dots on the right-hand side signify that an infinite number of similar terms are to be added.

$$e = 1 + \frac{1}{1} + \frac{1}{1 \cdot 2} + \frac{1}{1 \cdot 2 \cdot 3} + \frac{1}{1 \cdot 2 \cdot 3 \cdot 4} + \cdots$$

An infinite series always involves some rule that gives the value of every term. In the series above it is evident that each term is calculated by dividing the previous term by the number of the position of the term minus one. The value of e^x may be calculated by the series

$$e^x = 1 + \frac{x}{1} + \frac{x^2}{1 \cdot 2} + \frac{x^3}{1 \cdot 2 \cdot 3} + \frac{x^4}{1 \cdot 2 \cdot 3 \cdot 4} + \cdots$$

in which each term is found by dividing the preceding term by the number of the position of that term and multiplying it by x. The reader may want to verify that this series actually works by trying it, for example, for e^2. Series of this type occur so frequently that a special notation is used to describe them in a compact form. The following symbol means a sum of N terms in which the parameter n takes on successively the values of 1, 2, 3, up to N:

$$\sum_{n=1}^{N}$$

Thus, the two series presented above are described in this notation as

$$e = 1 + \sum_{n=1}^{\infty} \frac{1}{1 \cdot 2 \cdot 3 \cdot \cdots \cdot n}$$

$$e^x = 1 + \sum_{n=1}^{\infty} \frac{x^n}{1 \cdot 2 \cdot 3 \cdot \cdots \cdot n}$$

In general, there is no assurance that an infinite series actually has a unique value.

If the upper limit N is infinite, the sum of a series may be infinite or in some cases indeterminate. There is a variety of rules that can be used to establish whether or not a particular infinite series has a well-determined sum. However, since they would take us far afield, we say simply that the series that occur in this book all have well-determined sums.

Occasionally we wish to indicate a sum that includes only the terms for which n is odd or even. For these we use the notations

$$\sum_{0, 2, 4, 6, \ldots}^{\infty}$$

to represent an infinite number of terms beginning with $n = 0$ but including only those for which n is even, and

$$\sum_{1, 3, 5, 7, \ldots}^{\infty}$$

to represent an infinite number of terms including only those for which n is odd.

2. Fourier Series

The series described here are required only in the last section of this chapter. We list here some of the properties of the sine and cosine functions that will prove useful in this discussion:

$$\frac{d}{d\theta} \sin k\theta = k \cos k\theta, \qquad \frac{d}{d\theta} \cos k\theta = -k \sin k\theta \tag{8.1}$$

$$\int \sin k\theta \, d\theta = -\frac{1}{k} \cos k\theta, \qquad \int \cos k\theta \, d\theta = \frac{1}{k} \sin k\theta \tag{8.2}$$

$$\sin(-\theta) = -\sin \theta, \qquad \cos(-\theta) = \cos \theta \tag{8.3}$$

For n equal to an integer

$$\sin\frac{n\pi}{2} = 0 \quad \text{if } n \text{ even} \qquad \cos\frac{n\pi}{2} = (-1)^{n/2} \quad n \text{ even}$$
$$= (-1)^{(n-1)/2} \quad n \text{ odd} \qquad = 0 \quad n \text{ odd} \tag{8.4}$$

$$\sin n\pi = 0 \qquad \cos n\pi = (-1)^n \tag{8.5}$$

We now state without proof a remarkable theorem due to Fourier.

THEOREM. Given any function $f(x)$ which is defined between x equal to $-H$ and $+H$ and which is continuous in this interval or has a limited number of finite discontinuities in this interval, it is possible to express this function in terms of sine and cosine functions as

$$f(x) = \frac{a_0}{2} + \sum_{n=1}^{\infty} a_n \cos\frac{n\pi x}{H} + b_n \sin\frac{n\pi x}{H} \tag{8.6}$$

where the coefficients a_n and b_n are given by

$$a_n = \frac{1}{H}\int_{-H}^{H} f(x)\cos\frac{n\pi x}{H}\,dx \tag{8.7}$$

$$b_n = \frac{1}{H}\int_{-H}^{H} f(x)\sin\frac{n\pi x}{H}\,dx \tag{8.8}$$

Strictly speaking the series expansion is valid for $-H < x < H$. It may or may not be valid at the end points. See Problem 6 at the end of this chapter.

The proof of this theorem is difficult, but we can at least make it plausible by illustrating a few examples.

Example. Consider the function

$$f(x) = \frac{x}{H}$$

which consists of a ramp beginning with the value minus one at $-H$ and increasing to the right to a value of one at $+H$. The Fourier coefficients are given by

$$a_n = \frac{1}{H}\int_{-H}^{H} \frac{x}{H}\cos\frac{n\pi x}{H}\,dx$$

$$b_n = \frac{1}{H}\int_{-H}^{H} \frac{x}{H}\sin\frac{n\pi x}{H}\,dx$$

With the help of a set of integral tables these integrals are found to be

$$a_n = \frac{1}{H^2}\left[\left(\frac{H}{n\pi}\right)^2 \cos\frac{n\pi x}{H} + \left(\frac{H}{n\pi}\right)x \sin\frac{n\pi x}{H}\Big|_{-H}^{H}\right] = 0$$

$$b_n = \frac{1}{H^2}\left[\left(\frac{H}{n\pi}\right)^2 \sin\frac{n\pi x}{H} - \left(\frac{H}{n\pi}\right)x \cos\frac{n\pi x}{H}\Big|_{-H}^{H}\right]$$

$$= \frac{1}{H^2}\left(\frac{-H}{n\pi}\right)[H(-1)^n + H(-1)^n] = -\frac{2}{\pi}\frac{(-1)^n}{n}$$

Thus the Fourier series that represents the ramp function is

$$f(x) = -\frac{2}{\pi}\sum_{1,2,3,\ldots}^{\infty}\frac{(-1)^n}{n}\sin\frac{n\pi x}{H}$$

$$= +\frac{2}{\pi}\left[\sin\frac{\pi x}{H} - \frac{1}{2}\sin\frac{2\pi x}{H} + \frac{1}{3}\sin\frac{3\pi x}{H} + \cdots\right] \tag{8.9}$$

It is by no means obvious that this series really fits the prescribed function. Let us therefore plot the first three terms of the series and their sum and see if it comes close to the ramp (Figure 8-1).

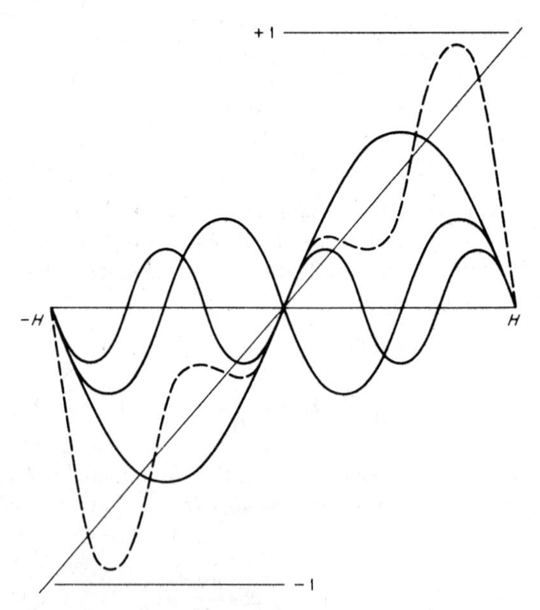

Figure 8-1. The approximation of the first three terms of a Fourier series to a ramp function. The three terms are shown individually as solid lines, as is the ramp function. The dotted line represents the sum of the three terms.

Now let us see if we can improve the approximation by plotting, for example, the sum of ten terms (Figure 8-2).

We see that at least for this function Fourier's theorem works.

The previous example was a continuous function. Now let us try a function with a break in it, a discontinuous function. Let

$$f(x) = \begin{cases} -1 & \text{for } x \text{ between } -H \text{ and } -H/2 \\ 1 & \text{for } x \text{ between } -H/2 \text{ and } +H/2 \\ -1 & \text{for } x \text{ between } H/2 \text{ and } H \end{cases}$$

which is a function that has a discontinuous step from -1 to 1 at $x = -H/2$ and another from $+1$ to -1 at $x = H/2$. Its Fourier coefficients are given by

$$a_n = \frac{1}{H}\int_{-H}^{-H/2} (-1)\cos\frac{n\pi x}{H}\,dx + \frac{1}{H}\int_{-H/2}^{H/2} (+1)\cos\frac{n\pi x}{H}\,dx + \frac{1}{H}\int_{H/2}^{H} (-1)\cos\frac{n\pi x}{H}\,dx$$

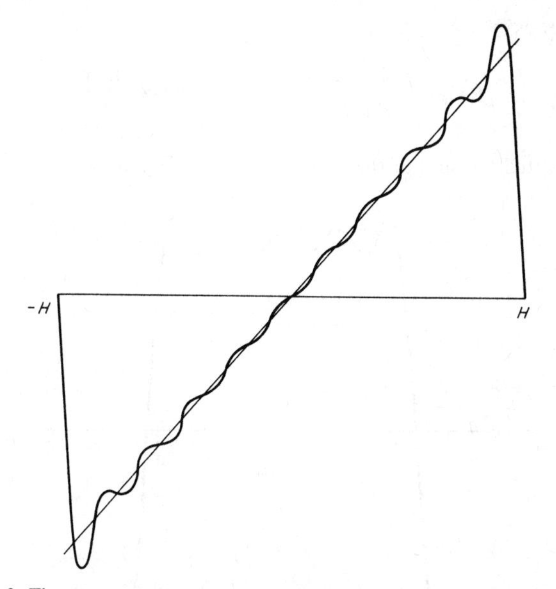

Figure 8-2. The representation of a ramp function by a Fourier series of ten terms. Not only is the approximation better over most of the range of the function, but it remains good into the corners close to the values $x = -H$ and $x = +H$, though it does not exactly fit the function at the extreme ends of the interval.

$$b_n = \frac{1}{H}\int_{-H}^{-H/2} (-1) \sin \frac{n\pi x}{H}\, dx + \frac{1}{H}\int_{-H/2}^{H/2} (+1) \sin \frac{n\pi x}{H}\, dx$$

$$+ \frac{1}{H}\int_{H/2}^{H} (-1) \sin \frac{n\pi x}{H}\, dx$$

where in both of the foregoing the integrals have been split as described in Section 1 of Chapter IV to incorporate the discontinuities. The integrals above are evaluated by making use of the relations at the beginning of this section; upon adding all the terms, we find

$$a_0 = 0$$

$$a_n = \frac{4}{n\pi}(-1)^{(n-1)/2}, \qquad n \text{ odd}$$

$$= 0, \qquad n \text{ even}$$

$$b_n = 0$$

Thus our series is

$$f(x) = +\frac{4}{\pi} \sum_{1, 3, 5, \ldots} \frac{1}{n}(-1)^{(n-1)/2} \cos \frac{n\pi x}{H} \tag{8.10}$$

We have plotted the sum of the first five terms in Figure 8-3.

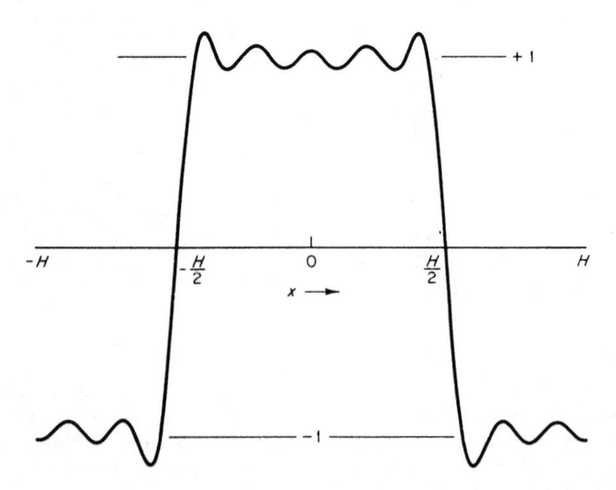

Figure 8-3. The approximation of a Fourier series of five terms to a rectangular function.

3. The Thermal Diffusion Equation

In this chapter we are going to introduce a new kind of differential equation, one that contains two different kinds of derivatives, a derivative with respect to a space dimension and a derivative with respect to time.

Consider a long, thin bar of metal, such as a curtain rod, different parts of which are at different temperatures. In order to describe the temperature along the length of such a bar, we must specify it as a function of two variables: x, the position along the bar; and t, the time. Let us derive a differential equation that describes the relationship between temperature changes as a function of x at a fixed time, and changes as a function of time for fixed x.

In order to do so, we must know something about the physics of the conduction of heat through a very thin layer. One finds experimentally that the rate at which heat will diffuse through a thin layer is given by

$$\frac{dQ}{dt} = \frac{K'A}{D}(T_1 - T_2) \tag{8.11}$$

where T_1 and T_2 are the temperatures on the two sides of the layer, A the cross-sectional area, D the thickness, and K' a constant called the thermal conductivity of the material, which has units of calories per degree per centimeter per second (cal deg^{-1} cm^{-1} sec^{-1}). Now let us consider the curtain rod broken up into a large number of such thin slices, each of a thickness Δx (Figure 8-4). By analogy with Eq. (8-11), the temperatures T_1 and T_2 are

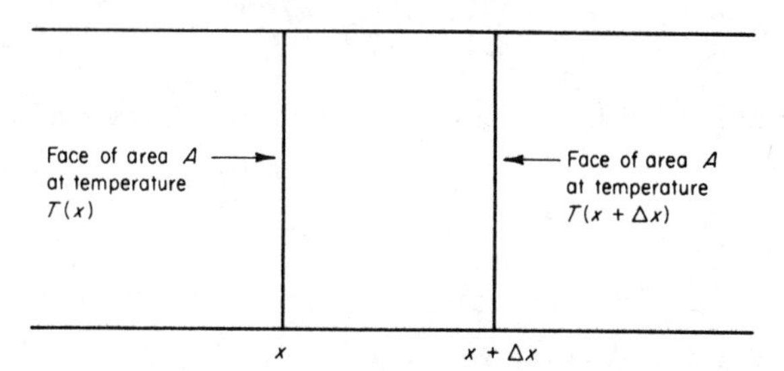

Figure 8-4. Illustration of the diffusion of heat through a very thin layer of metal.

replaced by

$$T_1 = T(x), \qquad T_2 = T(x + \Delta x)$$

and D is replaced by Δx. Thus,

$$\frac{dQ}{dt} = K'A\,\frac{T(x) - T(x + \Delta x)}{\Delta x} \tag{8.12}$$

We observe that, written in this form, the right-hand side of the equation is in the form of a derivative if Δx goes to zero:

$$\lim_{\Delta x \to 0} \frac{T(x + \Delta x) - T(x)}{\Delta x} = \frac{dT}{dx}$$

But notice that it is a different kind of derivative from the derivative on the left-hand side of Eq. (8.12). On the left-hand side of (8.12) we have the rate of flow of heat with respect to time at a fixed position x. On the right-hand side of Eq. (8.12) we have the rate of change of temperature with respect to position at a fixed time. Thus

$$\frac{dQ}{dt} = -K'A \frac{dT}{dx} \tag{8.13}$$

The quantity dT/dx is called the gradient of temperature. Heat "runs down" the gradient.

Equation (8.13) describes the rate at which heat travels through an infinitesimally thin slice of metal. We would like to know how the temperature of the metal within the slice changes, and in order to find this, we must ascertain how heat accumulates within a thin section. We therefore consider a small volume consisting of cross section A and thickness Δx, where this is now a new Δx, bounded on each side by slices such as are defined above (Figure 8.5). The rate at which heat enters the left-hand side of the volume is given by

$$\left.\frac{dQ}{dt}\right|_L = -K'A \frac{dT(x)}{dx} \tag{8.14}$$

while the rate at which it leaves on the right-hand side is given by

$$\left.\frac{dQ}{dt}\right|_R = -K'A \frac{dT(x + \Delta x)}{dx}$$

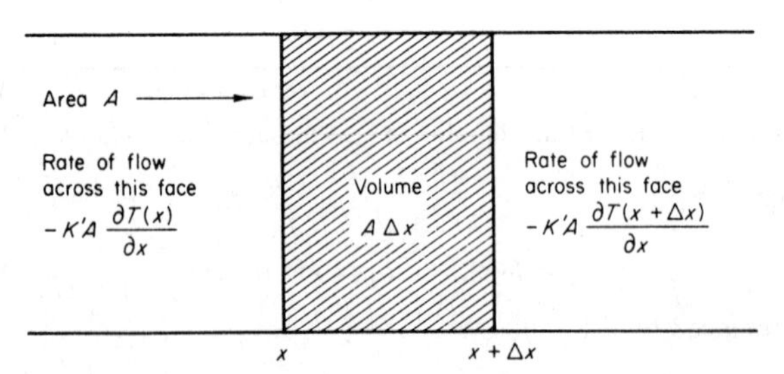

Figure 8-5. Illustrating the net input of heat to a section of thickness Δx based on the gradient of temperature at the slices that bound the section.

Notice that these two quantities are not necessarily equal because the derivatives with respect to x are evaluated at different points. The difference between these two quantities represents the rate at which heat accumulates within the volume. We know that a net change in heat content within a volume must produce a change in temperature, given by

$$\Delta T = \frac{\Delta Q}{\Omega}$$

where Ω is the heat capacity and is the product of the volume multiplied by the density multiplied by the specific heat. Therefore,

$$\Delta T = \frac{\Delta Q}{\rho\sigma A\,\Delta x}$$

where σ is the specific heat in calories per degree per gram (cal $\text{deg}^{-1}\ \text{gm}^{-1}$); ρ is the density; and $A\,\Delta x$ the volume. Thus, the rate of change of temperature is given by

$$\begin{aligned} \frac{dT}{dt} &= \frac{1}{A\rho\sigma\,\Delta x}\left(\left.\frac{dQ}{dt}\right|_L - \left.\frac{dQ}{dt}\right|_R\right) \\ &= \frac{K'A}{A\rho\sigma\,\Delta x}\left(\frac{dT(x+\Delta x)}{dx} - \frac{dT(x)}{dx}\right) \end{aligned} \tag{8.15}$$

which we see is now in the form of a derivative as $\Delta x \to 0$, but note that it is the derivative with respect to x of the derivative with respect to x. It is, therefore, a second derivative with respect to x:

$$\frac{dT}{dt} = K\frac{d^2T}{dx^2} \tag{8.16}$$

where $K = K'/\rho\sigma$. This is called the *diffusion equation in one dimension* (one space dimension).

This equation, along with a set of *starting* or *initial* conditions, the temperature at each point at time zero, and a set of *boundary conditions* that fix the temperature at the boundary of the region under consideration, determines the time course of temperature change at each point in the region for all future times.

Let us consider a long thin rod of uniform cross section whose ends are maintained at two different temperatures T_1 and T_2. Let us further assume that the ends have been held at these fixed temperatures long enough for the temperature at each point to stop changing with time. Then, at each point (each slice) the rate of change with respect to time is zero and the diffusion equation becomes

$$\frac{dT}{dt} = 0 = K\frac{d^2T}{dx^2}$$

for which the solution is

$$T = B_1 + B_2 x$$

The constants B_1 and B_2 are chosen to satisfy the given boundary conditions at $x = 0$ and $x = x_1$.

$$T(0) = B_1 = T_1, \qquad T(x_1) = B_1 + B_2 x_1 = T_2$$

Therefore

$$B_2 = \frac{1}{x_1}(T_2 - T_1)$$

and

$$T(x) = T_1 + (T_2 - T_1)\frac{x}{x_1}$$

We can verify that this is a solution by calculating the second derivative with respect to x and noting that the differential equation and the boundary conditions are satisfied:

$$\frac{dT}{dx} = \frac{T_2 - T_1}{x_1}, \qquad \frac{d^2T}{dx^2} = 0$$

$$T(0) = T_1, \qquad T(x_1) = T_1 + (T_2 - T_1)\frac{x_1}{x_1} = T_2$$

The problem just solved is called a *boundary value problem.* A boundary value problem consists of a differential equation, which describes how things change in infinitesimal volumes, and a set of conditions that apply at the boundaries, in this case the left and right end of the rod.

To better illustrate the physics of diffusion let us consider a considerably more complicated case, a long rod that initially has a higher temperature near the middle than it has on its ends, and see how this will change as a function of time (Figure 8-6). Notice that we have drawn the initial temperature so that at the ends of the rod it has zero slope, or zero derivative. This implies that heat is neither gained nor lost through the ends of the rod. To see this, consider the thin slices at the end of the rod. If the derivative with respect to x is zero at these points, we know from the heat diffusion equation (8.13) that no heat crosses these boundaries.

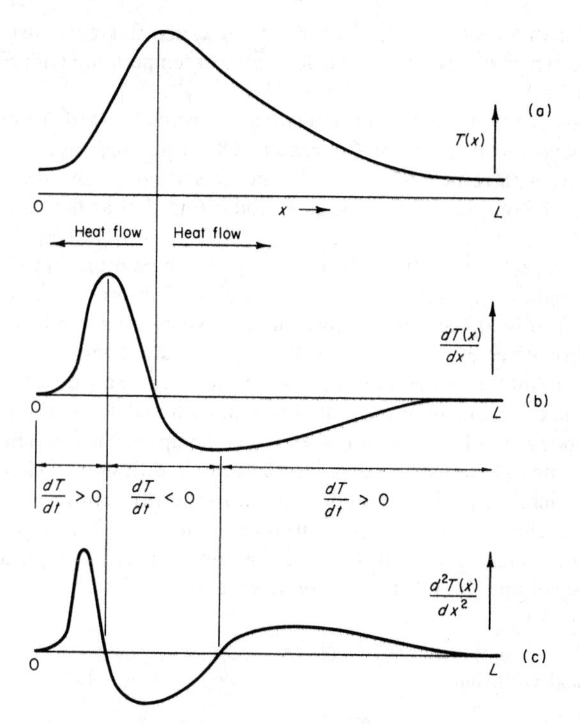

Figure 8-6. (a) Hypothetical initial temperature distribution along a rod of length L. (b) The rate of change of temperature as a function of x, along the same rod; the derivative dT/dx. (c) The second derivative of temperature as a function of x.

Now let us sketch in the first derivative of the temperature, which we do by plotting the slope of the temperature at each value of x (Figure 8.6b). On the left side of the rod, except at the very edge, the temperature slopes upward with increasing x. At the very edge, the slope (or gradient) is zero, since we assume that no heat enters or leaves through the ends. Near the middle of the rod, the slope is zero, since the temperature reaches a maximum at this point. On the right-hand side of the rod, the derivative is negative since the temperature is decreasing with increasing x until we reach the right-hand end of the rod, at which point the derivative is again zero.

Where the derivative is positive, heat flow is to the left, since

$$\frac{dQ}{dt} = -\frac{dT}{dx} K'A$$

Where the derivative is negative, heat flow is to the right.

The derivative dT/dx is itself a function of x, and therefore we can plot its derivative, which will be the second derivative of temperature with respect to x (Figure 8.6c).

We observe that the second derivative is positive toward the ends of the rod but negative just left of the center. Therefore, the rate of change of temperature with respect to time at this point is negative. On the other hand, at the ends of the rod dT/dt is positive, indicating that at the ends temperature increases with time.

As we expect, heat diffuses from the higher temperature in the middle of the rod toward the lower temperature at the ends. This process continues until the entire rod is at the same temperature, at which point both the first and second derivatives of temperature with respect to x are zero.

At this point the reader may begin to wonder why we are devoting so much attention to the problem of thermal diffusion. Although living things generate heat, the question of its diffusion does not come up very often. The reason is that the same equations also apply to the passive diffusion of chemical substances, which is a problem that does occur frequently in physiological literature. Let us therefore derive the diffusion equation for diffusion of chemical substances, which is similar but has the small added complication that within the volume the substance may be generated or lost.

4. Chemical Diffusion

The experimental law that governs chemical diffusion through a thin layer is analogous to that governing heat diffusion:

$$\frac{dQ}{dt} = \frac{KA}{D}(C_1 - C_2) \tag{8.17}$$

where concentrations C_1 and C_2 replace temperatures T_1 and T_2; Q is now in moles; K, A, and D are as before except that K now has dimensions of square centimeters per second ($\text{cm}^2\ \text{sec}^{-1}$).

We emphasize that this is an experimental result and cannot be logically derived. We further emphasize, as is particularly relevant in the biological case, that this diffusion is passive, involving no pumping of the substances and no fluid transport of it. Since the same basic law applies, the derivation of the diffusion equation proceeds in a fashion entirely analogous to that of heat flow and yields, for the net balance in an infinitesimal volume,

$$\left.\frac{dQ}{dt}\right|_{\text{net}} = \left.\frac{dQ}{dt}\right|_L - \left.\frac{dQ}{dt}\right|_R = KA\left[\frac{dC(x+\Delta x)}{dx} - \frac{dC(x)}{dx}\right] \tag{8.18}$$

which the reader should derive. Let us now add a term that represents the possible production or consumption of the substances in the volume. Here we have a variety of choices—most chemical substances are used at a rate proportional to their concentration. Thus, within the volume the net balance may include a term such as

$$\left.\frac{dQ}{dt}\right|_1 = -CL_1 A\,\Delta x$$

where L_1 has units of reciprocal seconds (sec^{-1}) and the subscript 1 means that L refers to consumption proportional to concentration. Note that $L_1 C$ has units of moles per second per cubic centimeter (moles $\text{sec}^{-1}\ \text{cm}^{-3}$). On the other hand, there is a singular exception to this rule, namely, oxygen in normal physiological ranges, which is used at a rate that is essentially independent of its concentration unless the concentration falls to anoxic levels. Furthermore, in the case of substances that are generated in the volume, their rate of generation is usually independent of their concentration. This is, however, also not a universal rule. The substance could be generated by a reversible reaction. For substances used at a constant rate, we add

$$\left.\frac{dQ}{dt}\right|_2 = -R_2 A\,\Delta x$$

where R_2 is in moles per second per cubic centimeter.

For substances generated by reversible reactions, for example,

$$S_1 + S_2 \underset{k_2}{\overset{k_1}{\rightleftharpoons}} P$$

$$\frac{dC_P}{dt} = S_1 S_2 k_1 - k_2 C_P$$

where S_1 and S_2 are the precursors of P and C_P is the concentration of P. Note that the left side of this equation is in terms of concentration to be consistent with the usual usage in kinetic theory. For uniformity, we convert this to

$$\left.\frac{dQ}{dt}\right|_3 = (S_1 S_2 k_1 - k_2 C_P) A\,\Delta x$$

In order to avoid having to investigate many combinations of circumstances, let us treat the most general case by allowing the substance to be used at a fixed rate per volume proportional to the constant R_2, which takes on negative values for generation at a fixed rate; let it be used in proportion to its concentration times the value of the constant L_1 and let it be generated from reversible reactions as determined by the values of k_1, k_2, S_1, S_2. Thus, each of our special cases is obtained by setting R_2, L_1, or k_1 and k_2 equal to zero. Adding these terms to the net balance equation, we find

$$\frac{dQ_{\text{net}}}{dt} = KA\left(\frac{dQ}{dt}\bigg|_L - \frac{dQ}{dt}\bigg|_R\right) + \frac{dQ}{dt}\bigg|_1 + \frac{dQ}{dt}\bigg|_2 + \frac{dQ}{dt}\bigg|_3$$

In the case of heat, the effect of a nonzero net flow was to change the temperature of the volume. The effect of nonzero chemical flow is to change the concentration

Since

$$C = \frac{Q}{V} = \frac{Q}{A\,\Delta x}$$

$$\frac{dC}{dt} = \frac{1}{A\,\Delta x}\frac{dQ}{dt}\bigg|_{\text{net}}$$

$$= \frac{K}{\Delta x}\left(\frac{dC(x+\Delta x)}{dx} - \frac{dC(x)}{dx}\right) - L_1 C - R_2 + S_1 S_2 k_1 - k_2 C$$

which in the limit as Δx goes to zero yields

$$\frac{dC}{dt} = K\frac{d^2C}{dx^2} - L_1 C - R_2 + S_1 S_2 k_1 - k_2 C$$

Let us now combine some of the terms, using a notation analogous to that of Chapter V. Let

$$R = R_2 - S_1 S_2 k_1, \qquad L = L_1 + k_2$$

Thus

$$\frac{dC}{dt} = K\frac{d^2C}{dx^2} - R - LC \tag{8.19}$$

The foregoing derivation is based on long thin rods. There is nothing, however, that restricts it to a thin rod provided we are assured that diffusion occurs in only one direction. This will be the case for uniform volumes in which the lateral dimensions (perpendicular to the direction of diffusion) are large compared to the diffusing depth of the substance in question and provided that the surface is at a uniform concentration at all points. This is perhaps made clear by considering a counterexample. Suppose one has a large cube of metal, say 6 cm on a side, and that one touches the middle of one face with a hot soldering iron. In this case, diffusion would not be in one direction but would occur radially from the point of contact of the soldering iron. On the other hand, if one face of the cube is simply placed against a hot surface of uniform temperature, diffusion is essentially along parallel lines perpendicular to the hot surface.

The mathematical justification for considering only one dimension of a three-dimensional object is clear when one considers the derivation of the diffusion equation. The thin slices of the curtain rod are now replaced by similar thin slices of a broad section, and as long as the heat flow or chemical flow is primarily perpendicular to these thin slices, the derivation proceeds in exactly the same way.

In general, solutions of the time-dependent diffusion equation are fairly complicated. We will, therefore, initially consider a few special cases of the steady-state diffusion equation for mixed concentration-dependent and concentration-independent loss or generation of the diffusing substance. But before doing so we must express a word of caution about concentration-independent loss. We must always formulate problems in a way that avoids a solution that leads to a negative concentration. This is physically unacceptable. What actually happens, of course, is that the assumption of concentration independence is violated when the concentration becomes too small. To avoid possible negative values, we shall always formulate problems in a way that assures that the solution cannot go negative. From this point on, unless otherwise stated, we are considering steady-state conditions for which

$$\frac{dC}{dt} = 0$$

and, therefore, the steady-state one-dimensional diffusion equation for mixed concentration-dependent and concentration-independent loss becomes

$$\frac{d^2C}{dx^2} - \frac{L}{K}C = \frac{R}{K} \tag{8.20}$$

We are going to investigate the following steady-state cases.

1. A very thick (semi-infinite) section with mixed loss.
2. Finite section with zero concentration and flow at one face and mixed loss.
3. Same as case 2 with only concentration-independent loss (see Problem 2).
4. Same as case 2 with only concentration-dependent loss (see Problem 3).
5. Semi-infinite section with fixed concentration at the face and diffusing substance generated internally at a fixed rate.
6. Semi-infinite block with mixed loss above a threshold concentration and concentration-dependent loss below the threshold.
7. Section of finite thickness $2D$ with fixed rate of generation and specified surface concentration.

8. Section of thickness $2D$ with concentration-dependent loss (see Problem 5 at end of chapter).
9. Spherical diffusion with mixed loss and concentration given on surface.
10. Cylindrical diffusion with concentration-independent loss (the Krogh equation).

5. Semi-Infinite Section with Mixed Loss

Let us consider first a semi-infinite block extending from x equal to zero to x equal to infinity. To avoid the possibility that the result will imply a negative concentration, we shall assume that the concentration is zero at x equal to zero, all flow of the diffusing substance is from right to left, and no flow occurs out of the left face. We are thus assured that all computed concentrations to the right of x equal to zero are positive and that therefore the assumed condition of fixed concentration-independent loss will not be violated. (Note that we are implying here that the concentration-independent loss occurs down to zero concentration.) The differential equation (8.20) and accessory conditions to be satisfied are

$$\frac{d^2C}{dx^2} - \frac{L}{K}C = \frac{R}{K}, \qquad 0 < x < \infty$$

$$C(0) = 0, \qquad \frac{dC(0)}{dx} = 0 \tag{8.21}$$

Since x extends from 0 to infinity, this problem is well suited for a solution by means of the Laplace transform. Transforming the differential equation yields

$$s^2 g(s) - sC(0) - \frac{dC(0)}{dx} - \frac{L}{K} g(s) = \frac{R}{sK}$$

which becomes, upon inserting the accessory conditions,

$$g = \frac{R}{Ks(s^2 - L/K)}$$

Thus from entry 17 (Appendix II) setting a^2 equal to L/K, we get

$$C(x) = \frac{R}{2L}(e^{x\sqrt{L/K}} + e^{-x\sqrt{L/K}} - 2) \tag{8.22}$$

We can find the flow past a plane at x_1 from the derivative of C with respect to x evaluated at x_1:

$$F(x_1) = -KA\frac{dC(x_1)}{dx} = -\frac{AKR}{2L}\sqrt{\frac{L}{K}}\left(e^{x_1\sqrt{L/K}} - e^{-x_1\sqrt{L/K}}\right)$$

$$= -\frac{AR}{2}\sqrt{\frac{K}{L}}\left(e^{x_1\sqrt{L/K}} - e^{-x_1\sqrt{L/K}}\right) \tag{8.23}$$

where A is the cross-sectional area of the plane. The minus sign arises because the flow is in the negative direction.

6. Finite Section with Mixed Loss

We can easily reduce the solution (8.22) to that for a finite section by cutting it off at x_1. Thus a section that extends from x equal to zero to x_1 and has zero concentration and flow at the left edge has the solution shown in Figure 8-7.

The way in which we have stated this problem is somewhat different from the way in which it would usually arise. Usually we are given a concentration and asked to determine the depth of penetration of the diffusing substance.

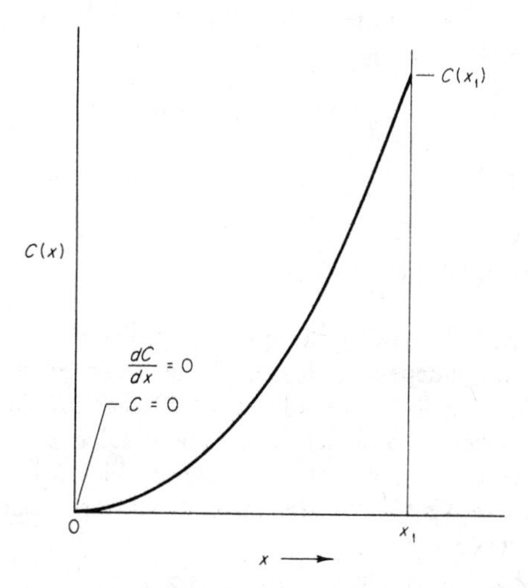

Figure 8-7. The concentration in a semi-infinite medium in which the diffusing substance is lost by both concentration-dependent and concentration-independent loss, and in which the concentration at the left-hand boundary of the section and the flow of the diffusing substance across this boundary are both zero.

Within the restrictions on this solution (which we shall remove later) that the penetration stops at some point x equal to zero, for a given concentration at x_1 we can find the depth of penetration by solving

$$C(x_1) = \frac{R}{2L}(e^{x_1\sqrt{L/K}} + e^{-x_1\sqrt{L/K}} - 2) \tag{8.24}$$

for x_1 numerically.

We should take note of two special cases. In the first L is zero; for this case

$$C(x) = \frac{Rx^2}{2K} \tag{8.25}$$

$$F(x_1) = -KA\frac{dC(x)}{dx} = -RAx_1$$

The second special case is that of R equal to zero. In this case C cannot become zero for any finite depth of penetration into the section because the rate of loss is proportional to C. Equation (8.21) is valid but its accessory conditions are not. On the other hand, there is no danger of calculating a negative value of C since it cannot become zero. The technique of solving this problem is to consider the left face (x zero) of an infinite block to be at concentration $C(0)$ and to require that C remain finite for all positive x. The solution, which is left for the reader to derive, is

$$C(x) = C(0)e^{-x\sqrt{L/K}} \tag{8.26}$$

(See Problems 2 and 3.)

7. Threshold-Dependent Loss

Suppose we wish to modify the previous problem to cover the case in which concentration-independent loss does not occur below some threshold concentration C_T. To do so we solve two different problems. We assume that C_T exists at x zero. To the left of x zero we solve for C when R equals zero. To the right of x zero we solve the case in which R is not zero and the concentration at x zero is C_T. We then match the flow rate calculated from the two solutions across x zero.

To the left of x zero we have

$$\frac{d^2C}{dx^2} - \frac{L}{K}C = 0, \qquad C(0) = C_T$$

Equation (8.26) suggests that we try a solution of the form

$$C(x) = C_T e^{x\sqrt{L/K}} \tag{8.27}$$

where we have chosen a positive sign in the exponent because the solution must go to zero for large negative values of x. The reader may verify that this is a solution to the foregoing differential equation. We find the flow past x zero from the above from

$$F(0) = -KA\frac{dC(0)}{dx} = -KA\sqrt{\frac{L}{K}}C_T \tag{8.28}$$

To the right of x zero we have

$$\frac{d^2C'}{dx^2} - \frac{L}{K}C' = \frac{R}{K}$$

for which we try as a solution

$$C'(x) = -\frac{R}{L} + B_2 e^{x\sqrt{L/K}} + B_3 e^{-x\sqrt{L/K}}$$

and now fix the constants by matching the flow and concentration at $x = 0$ to (8.27) and (8.28)

$$C'(0) = C(0) = -\frac{R}{L} + B_2 + B_3$$

$$F(0) = -KAC(0)\sqrt{\frac{L}{K}} = -KA\frac{dC'(0)}{dx}$$

$$= -KA\left(B_2\sqrt{\frac{L}{K}} - B_3\sqrt{\frac{L}{K}}\right)$$

From this pair of equations we find

$$B_2 = C(0) + \frac{R}{2L}, \qquad B_3 = \frac{R}{2L}$$

$$C'(x) = C(0)e^{x\sqrt{L/K}} + \frac{R}{2L}(e^{x\sqrt{L/K}} + e^{-x\sqrt{L/K}}) - \frac{R}{L} \tag{8.29}$$

Normally we are given the surface concentration $C(x_1)$ and $C(0)$, the concentration below which concentration-independent loss does not occur, and from these we usually want to find the depth of penetration of concentration-independent loss. Once again we do this by regarding x_1 as the unknown and solving

$$C'(x_1) = \left(C(0) + \frac{R}{2L}\right)e^{x_1\sqrt{L/K}} + \frac{R}{2L}e^{-x_1\sqrt{L/K}} - \frac{R}{L} \tag{8.30}$$

for x_1 numerically. A typical solution resembles the diagram in Figure 8-8.

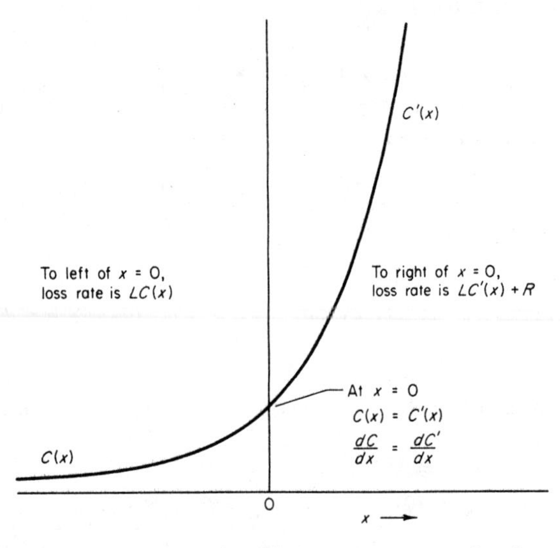

Figure 8-8. The concentration of a diffusing substance as a function of distance x, where concentration-independent loss occurs only above a threshold C_T but concentration-dependent loss occurs throughout the section.

8. Semi-Infinite Section with Internal Generation

Now let us consider the case in which the substance is generated within the volume and flows out of the left edge, which is at concentration $C(0)$. In this case R is negative, but to simplify the interpretation let us replace $-R$ by I. Thus the differential equation and accessory conditions to be satisfied are

$$\frac{d^2C}{dx^2} - \frac{L}{K}C = -\frac{I}{K} \tag{8.31}$$

where I is positive and $C(0)$ is given. Let us try as a solution to this problem the following:

$$C(x) = \frac{I}{L} + B_1 e^{x\sqrt{L/K}} + B_2 e^{-x\sqrt{L/K}}$$

Now if L is not zero, the concentration cannot become greater than

$$C_{max} = \frac{I}{L}$$

Therefore, B_1 must be zero, since the exponential term associated with it goes to infinity for large x. Thus we propose the following solution, which we proceed to verify,

$$C(x) = \frac{I}{L} + B_2 e^{-x\sqrt{L/K}}$$

We determine B_2 from the condition on $C(0)$ at $x = 0$:

$$C(0) = \frac{I}{L} + B_2$$

Thus

$$C(x) = \frac{I}{L} + \left(C(0) - \frac{I}{L}\right)e^{-x\sqrt{L/K}} \tag{8.32}$$

To verify, we calculate the second derivative of C and add $-LC/K$:

$$\frac{d^2C}{dx^2} = \frac{L}{K}\left(C(0) - \frac{I}{L}\right)e^{-x\sqrt{L/K}}$$

$$\frac{L}{K}C = \frac{I}{K} + \left(\frac{LC(0)}{K} - \frac{I}{K}\right)e^{-x\sqrt{L/K}}$$

Therefore

$$\frac{d^2C}{dx^2} - \frac{L}{K}C = -\frac{I}{K}$$

Thus the given differential equation is satisfied and since at $x = 0$

$$C(0) = \frac{I}{L} + \left(C(0) - \frac{I}{L}\right) e^0 = C(0)$$

the accessory condition is also satisfied.

9. Finite Section with Internal Generation

Now let us try the same problem for a section of finite thickness. In this case it is convenient to let the section be from x equal to $-D$ to x equal to $+D$

for reasons that will be obvious. Let the concentration on both sides of the section be $C(D)$. Once again we try a solution of the form

$$C(x) = \frac{I}{L} + B_1 e^{x\sqrt{L/K}} + B_2 e^{-x\sqrt{L/K}} \tag{8.33}$$

$$C(D) = C(-D) \qquad \text{given}$$

Since the two exponential terms exchange values when we substitute $-D$ for D, B_1 must equal B_2

$$C(x) = \frac{I}{L} + B(e^{x\sqrt{L/K}} + e^{-x\sqrt{L/K}})$$

and B is found given

$$C(D) = \frac{I}{L} + B(e^{D\sqrt{L/K}} + e^{-D\sqrt{L/K}})$$

Thus

$$C(x) = \frac{I}{L} + \left(C(D) - \frac{I}{L}\right) \frac{e^{x\sqrt{L/K}} + e^{-x\sqrt{L/K}}}{e^{D\sqrt{L/K}} + e^{-D\sqrt{L/K}}} \tag{8.34}$$

and the flow through each surface is

$$F(D) = -KA \frac{dC(D)}{dx}$$

$$= KA\left(C(D) - \frac{I}{L}\right) \sqrt{\frac{L}{K}} \frac{e^{D\sqrt{L/K}} - e^{-D\sqrt{L/K}}}{e^{D\sqrt{L/K}} + e^{-D\sqrt{L/K}}} \tag{8.35}$$

We leave the two special cases of $L = 0$ and $I = 0$ as exercises.

10. Spherical Diffusion

There is a completely different kind of one-dimensional problem that is actually a problem in three spatial dimensions but can be reduced to a single space dimension. That is the problem of spherical diffusion. Let us consider a sphere out of which a substance is diffusing. In this case we must adopt a slightly different strategy. Instead of considering thin slices, we consider thin spherical shells of thickness Δr and a radius r from which we shall build up a sphere. We can find the rate at which a diffusing substance will pass through the surface of such a shell by considering a sufficiently small area of it, so that the curvature within this area will be negligible. One can visualize this, for example, as a postage stamp stuck to the surface of a basketball. For an area small enough so that its curvature is negligible, the rate at which a diffusing substance will cross this area is exactly analogous to the rate at which it would cross the slices of the curtain rod example,

$$\frac{dQ}{dt} = -KA\frac{dC(r)}{dr}$$

where A is the area of the postage stamp.

Now we can proceed to cover the entire surface of the basketball with postage stamps, so that the area term becomes the area of a sphere of radius r, which is $4\pi r^2$. Thus, the rate at which the diffusing substance diffuses outward through a spherical shell is given by

$$\frac{dQ}{dt} = -4\pi K r^2 \frac{dC(r)}{dr}$$

Now let us consider the net balance for a shell of thickness Δr and, as before, we allow the option that the substance may be consumed or generated within the volume of the shell where the volume is given by

$$V_{\text{shell}} = 4\pi r^2 \, \Delta r$$

In the case of the curtain rod, when we did the net balance calculation, the area A on both sides of the slice was the same. In the case of the spherical shell of thickness Δr, its inside area is

$$A_{\text{in}} = 4\pi r^2$$

Its outside area, however, is slightly greater and is given by

$$A_{\text{out}} = 4\pi(r + \Delta r)^2$$

Therefore, the net balance equation must be written in the form

$$\left.\frac{dQ}{dt}\right|_{\text{net}} = 4\pi K\left(-r^2\frac{dC(r)}{dr} + (r + \Delta r)^2\frac{dC(r + \Delta r)}{dr}\right)$$
$$- 4\pi r^2 \, \Delta r(L_1 C + R_2 + k_2 C - S_1 S_2 k_1)$$

where, as in the previous example, the second bracket represents the generation or loss of the diffusing substance per unit volume, and $4\pi r^2 \, \Delta r$ is the volume of the shell. Now, in order to make the next step clear, we define a function

$$W(r) = +r^2\frac{dC(r)}{dr}$$

and thus

$$W(r + \Delta r) = + (r + \Delta r)^2 \frac{dC(r + \Delta r)}{dr}$$

The net balance equation becomes

$$\left.\frac{dQ}{dt}\right|_{\text{net}} = 4\pi K(W(r + \Delta r) - W(r)) - 4\pi r^2 \,\Delta r(L_1 C + R_2 + k_2 C - S_1 S_2 k_1)$$

and, dividing through by $4\pi \,\Delta r \, r^2$ to get the rate of concentration change in the shell, we get

$$\frac{dC}{dt} = \frac{1}{4\pi r^2 \,\Delta r} \left.\frac{dQ}{dt}\right|_{\text{net}} = \frac{K}{r^2} \frac{W(r + \Delta r) - W(r)}{\Delta r} + (S_1 S_2 k_1 - k_2 C - R_2 - CL_1)$$

We then take the limit as Δr goes to zero and find

$$\frac{dC}{dt} = \frac{K}{r^2} \frac{dW}{dr} - R - LC$$

where L and R are as previously defined.

Since

$$W = +r^2 \frac{dC}{dr}, \qquad \frac{dW}{dr} = + \frac{d}{dr}\left(r^2 \frac{dC}{dr}\right)$$

therefore

$$\frac{dC}{dt} = \frac{K}{r^2} \frac{d}{dr}\left(r^2 \frac{dC}{dr}\right) - R - LC \tag{8.36}$$

This is the diffusion equation applied to a symmetric sphere; once again, let us solve for the steady state in which $dC/dt = 0$. This particular solution is based on a trick, which is the relationship

$$\frac{1}{r}\frac{d^2(rC)}{dr^2} = \frac{1}{r}\frac{d}{dr}\left(C + r\frac{dC}{dr}\right) = \frac{1}{r}\left(r\frac{d^2C}{dr^2} + 2\frac{dC}{dr}\right)$$

$$= \frac{d^2C}{dr^2} + \frac{2}{r}\frac{dC}{dr} = \frac{1}{r^2}\frac{d}{dr}\left(r^2 \frac{dC}{dr}\right)$$

Therefore, letting $P = rC$ in the steady state, we have

$$\frac{d^2P}{dr^2} - \frac{LP}{K} = r\frac{R}{K} \tag{8.37}$$

which has now transformed the spherical diffusion equation into a form similar to the linear diffusion equation. We now, therefore, try trial solutions in the form

$$P = B_1 e^{r\sqrt{L/K}} + B_2 e^{-r\sqrt{L/K}} - \frac{rR}{L}$$

or

$$C = \frac{B_1}{r} e^{r\sqrt{L/K}} + \frac{B_2}{r} e^{-r\sqrt{L/K}} - \frac{R}{L} \tag{8.38}$$

which, we find by differentiating, will satisfy the spherical diffusion equation. However, we observe that written in this form there is the possibility that the concentration becomes infinite at the center of the sphere where r is equal to 0 and the only way we can avoid this is to choose $B_1 = -B_2$. Thus, the solution to the spherical diffusion equation in the steady state is

$$C(r) = B\left(\frac{e^{r\sqrt{L/K}} - e^{-r\sqrt{L/K}}}{r}\right) - \frac{R}{L} \tag{8.39}$$

where B must be adjusted to fit the concentration or rate of flow at the surface of the sphere defined by $r = r_0$. This is not the most general case of diffusion in a sphere. It assumes the diffusion is purely radial. This would not be the case, for example, if we were boiling an egg and only half the egg were submerged in the boiling water. But the spherically symmetric case is a useful one since it is simple and since it is a rough approximation for some single-cell phenomena.

If $C(r_0)$ is given,

$$B = \frac{r_0}{e^{r_0\sqrt{L/K}} - e^{-r_0\sqrt{L/K}}}\left(C(r_0) + \frac{R}{L}\right) \tag{8.40}$$

See Problem 9.

11. Cylindrical Diffusion and the Krogh Equation

The problem of diffusion of oxygen around capillaries is obviously one of great importance to physiologists. A rigorous solution is quite difficult, but one can find approximate solutions by taking advantage of the fact that a capillary is relatively long compared to its diameter and to the distance between capillaries. Thus, an appropriate model for a short length of a capillary is that of a cylinder in which it is assumed that there is essentially no variation along the length of the cylinder but only radially from it. This is obviously not a good model of a capillary, since it is well known that the partial pressure of oxygen changes a great deal over the full length of a capillary, but one can incorporate this effect by putting end to end short cylinders each of which is short enough so that the variation along its length can be neglected.

The logic follows closely that of the case of spherical diffusion. One builds up a volume out of concentric cylindrical shells, each of length H, radius r, and thickness Δr (Figure 8-9). One then says that the inside of such a cylinder

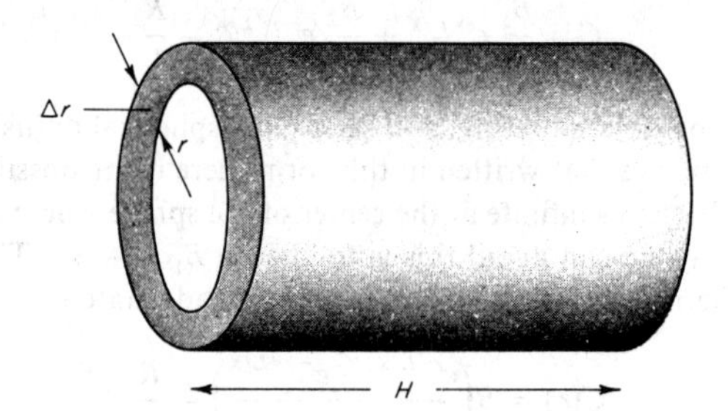

Figure 8-9. Model for cylindrical diffusion.

can be approximated as small flat "postage stamp" areas, and that the law of diffusion for a small flat area applies to each of these:

$$\frac{dQ}{dt} = -KA'\frac{dC}{dr}$$

where A' is the area of each postage stamp. One then describes the diffusion into the cylindrical shell as

$$\left.\frac{dQ}{dt}\right|_{\text{in}} = -K2\pi rH\frac{dC(r)}{dr}$$

where $2\pi rH$ is the area of the inside of the shell. Diffusion out of the shell is

$$\left.\frac{dQ}{dt}\right|_{\text{out}} = -K2\pi(r+\Delta r)H\frac{dC(r+\Delta r)}{dr}$$

where $2\pi H(r + \Delta r)$ is the area of the outside. Within the shell oxygen is consumed at a rate R per unit volume per second, and since the volume of the shell is

$$2\pi r\,\Delta r\,H$$

we have the net balance equation

$$\left.\frac{dQ}{dt}\right|_{\text{net}} = 2\pi KH\left[(r+\Delta r)\frac{dC(r+\Delta r)}{dr} - r\frac{dC(r)}{dr}\right] - 2\pi r\,\Delta r\,HR$$

Also

$$\left.\frac{dQ}{dt}\right|_{\text{net}} = 2\pi r\, \Delta r\, H \frac{dC}{dt}$$

therefore

$$\frac{dC}{dt} = K \frac{(r + \Delta r)\dfrac{dC(r + \Delta r)}{dr} - r\dfrac{dC}{dr}}{r\, \Delta r} - R$$

$$= \frac{K}{r}\frac{d}{dr}\left(r\frac{dC}{dr}\right) - R$$

In the steady state dC/dt is zero and we have

$$\frac{d}{dr}\left(r\frac{dC}{dr}\right) = \frac{Rr}{K}$$

This is called the *Krogh equation* [it is usually given in terms of partial pressures rather than concentration (11)]. It is easily integrated once to yield

$$r\frac{dC}{dr} = \frac{Rr^2}{2K} + B_1$$

where B_1 is an arbitrary constant or

$$\frac{dC}{dr} = \frac{Rr}{2K} + \frac{B_1}{r}$$

A second integration using the relation

$$\int \frac{1}{r}\, dr = \log r$$

(Appendix 1) yields

$$C = \frac{Rr^2}{4K} + B_1 \log r + B_2$$

where B_2 is a second arbitrary constant.

As usual, the arbitrary constants are chosen to fit accessory conditions of the physical problem. In this case we assume that we know the value of C at the surface of the capillary r_1. The second condition (we need two because there are two arbitrary constants to be fitted) is a bit more tenuous. It is assumed that there is some radius beyond which diffusion does not occur.

More accurately, at this radius, diffusion comes from adjacent capillaries. The condition then is

$$\frac{dC(r_2)}{dr} = 0$$

where r_2 is approximately half the mean distance between capillaries.

The easy way to evaluate the constants is to write C' as the difference between $C(r)$ and $C(r_1)$

$$C' = C(r) - C(r_1) = \frac{R(r^2 - r_1^{\,2})}{4K} + B_1(\log r - \log r_1)$$

Now set dC'/dr at r_2 equal to zero

$$\frac{dC'}{dr} = \frac{Rr_2}{2K} + \frac{B_1}{r_2} = 0, \qquad B_1 = -\frac{Rr_2^{\,2}}{2K}$$

Noting that

$$\log r - \log r_1 = \log \frac{r}{r_1}$$

we have

$$C(r) = \frac{R}{2K}\left(\frac{r^2 - r_1^{\,2}}{2} - r_2^{\,2} \log \frac{r}{r_1}\right) + C(r_1)$$

Although the differential equation derived above is correct within the given assumptions, the result must not be taken too seriously. In particular, the assumption that lengthwise effects can be neglected is probably a poor one. It is also probable that fluid motion (as opposed to diffusion) plays an important part in oxygen transport. There is an extensive literature on this subject; see (11).

12. Time-Dependent Diffusion

We now demonstrate solutions to the diffusion equation in the time-dependent or nonsteady-state case. There exists a great body of mathematical literature on the solution of equations of this type and it is among the most elegant of applied mathematics. Unfortunately, it is also very complicated; in fact, most practical problems are solved numerically. Nevertheless, one can acquire a great deal of understanding of the nature of the diffusion process by studying *two typical cases.* We shall therefore describe these problems, indicate their solutions, and verify the solutions by showing that they do indeed satisfy the differential equations and that they can be made at the initial time to agree with the stated initial conditions of the problem.

The time-dependent diffusion equation usually arises in a problem in which the condition of the system, for example, the temperature distribution or the concentration distribution, is specified at time zero and the task at hand is to find out how this concentration or temperature changes after time zero as a result of the diffusion process.

Consider the following problem. A uniform bar that extends from x equal to $-D$ to x equal to D is initially at temperature zero. At time zero the ends of the bar are brought to temperature T_1, and remain at this temperature thereafter. How does the temperature vary with time along the length of the bar? The reader should recognize that this is exactly the same as a tissue slice, initially having zero concentration of a diffusible substance, that is immersed in a bath whose concentration of the substance is C_1 provided that the substance is not consumed within the volume.

From Eq. (8.16) we find that the differential equation to be satisfied is

$$\frac{dT}{dt} = K\frac{d^2T}{dx^2}$$

with the boundary conditions

$$T(-D, t) = T(D, t) = T_1 \qquad \text{for all times } t \text{ after zero}$$

and the initial condition to be satisfied is

$$T(x, 0) = 0 \qquad \text{for all } x \text{ from } -D \text{ to } D \text{ at time zero}$$

The solution is given by

$$T(x, t) = T_1\left[1 - \frac{4}{\pi}\sum_{1, 3, 5, \ldots}^{\infty} \frac{(-1)^{(n-1)/2}}{n} e^{-(n\pi/2D)^2Kt} \cos\frac{n\pi x}{2D}\right] \tag{8.41}$$

which we shall verify by first calculating its derivatives with respect to x and t.

To find the derivative of this solution with respect to t we notice that each term of the infinite series contains a factor

$$e^{-(n\pi/2D)^2Kt}$$

and that t occurs in these factors and nowhere else in the solution. Therefore, for the purpose of differentiating with respect to t all other terms of the expression (8.41) may be regarded as constants. We find the derivative of each term containing t from the relation

$$\frac{d}{dt} e^{-(n\pi/2D)^2Kt} = -\left(\frac{n\pi}{2D}\right)^2 Ke^{-(n\pi/2D)^2Kt}$$

and therefore, the derivative of (8.41) is given by

$$\frac{dT(x, t)}{dt} = T_1 \frac{4}{\pi} \sum_{1, 3, 5,}^{\infty} \frac{(-1)^{(n-1)/2}}{n} \left(\frac{n\pi}{2D}\right)^2 Ke^{-(n\pi/2D)^2Kt} \cos \frac{n\pi x}{2D}$$

To find the second derivative with respect to x we note that each term of the infinite series contains a factor

$$\cos \frac{n\pi x}{2D}$$

and that x occurs only in these factors. Their second derivatives are found as follows.

$$\frac{d}{dx} \cos \frac{n\pi x}{2D} = -\frac{n\pi}{2D} \sin \frac{n\pi x}{2D}$$

$$\frac{d^2}{dx^2} \cos \frac{n\pi x}{2D} = -\left(\frac{n\pi}{2D}\right)^2 \cos \frac{n\pi x}{2D}$$

Thus, the second derivative of (8.41) is

$$\frac{d^2T}{dx^2} = T_1 \frac{4}{\pi} \sum_{1, 3, 5, \ldots}^{\infty} \frac{(-1)^{(n-1)/2}}{n} e^{-(n\pi/2D)^2Kt} \left(\frac{n\pi}{2D}\right)^2 \cos \frac{n\pi x}{2D}$$

We now note that if the series that is the second derivative with respect to x is multiplied by K, it is identical to the series that is the derivative of (8.41) with respect to time, and that, therefore, (8.16) is satisfied. We then note that at x equal to $-D$ and x equal to D all of the cosine terms are zero, and the boundary condition is therefore satisfied.

To establish that the initial condition is satisfied, we evaluate $T(x, 0)$, which is

$$T(x, 0) = T_1 \left[1 - \frac{4}{\pi} \sum_{1, 3, 5, \ldots}^{\infty} \frac{(-1)^{(n-1)/2}}{n} \cos \frac{n\pi x}{2D}\right]$$

and we note that if $2D$ equals H, the infinite sum is the same as that calculated in Section 1 for a function $f(x)$, which is

$$\begin{aligned} f(x) &= -1, \qquad -H < x < -H/2 \\ f(x) &= 1, \qquad -H/2 < x < H/2 \\ f(x) &= -1, \qquad H/2 < x < H \end{aligned}$$

Therefore one minus the infinite sum will be zero from $-D$ to D, which is the required initial condition. Note that $T(x, 0)$ thus calculated is not zero outside of the range $-D$ to D; it has the value 2, but this is irrelevant. The initial condition is satisfied over the range in question in this problem and the boundary conditions are satisfied at all times.

Now let us demonstrate the solution for diffusion into a sphere. From (8.36), setting R and L to zero, we have for thermal diffusion

$$\frac{dT}{dt} = \frac{K}{r^2}\frac{d}{dr}\left(r^2 \frac{dT}{dr}\right) \tag{8.42}$$

If the initial temperature is zero

$$T(r, 0) = 0$$

and the surface temperature is

$$T(r_0, t) = T_1$$

we find

$$T(r, t) = T_1\left[1 - \frac{2r_0}{\pi r} \sum_{1, 2, 3, 4, \ldots}^{\infty} \frac{(-1)^{n-1}}{n} e^{-(n\pi/r_0)^2 Kt} \sin \frac{n\pi r}{r_0}\right] \tag{8.43}$$

verification of which we leave as an exercise.

Let us examine in detail the nature of these solutions. In both cases they consist of T_1 multiplied by one minus an infinite sum. Each term of the infinite sum decays with time until at infinite time only the T_1 remains. The rate at which each term decays is e^{-n^2At} (where A is a constant that is the same in all terms) which is faster for larger n than for smaller n. The first term of the infinite series decays more slowly than any of the succeeding terms. It is also initially the largest term, since each term contains a factor of $1/n$. Thus, the initial temperature profile, which is rectangular (zero except at the ends of the bar or surface of the sphere), rounds off its corners very quickly and then slowly rises to a uniform value T_1. Figure 8-10 shows how

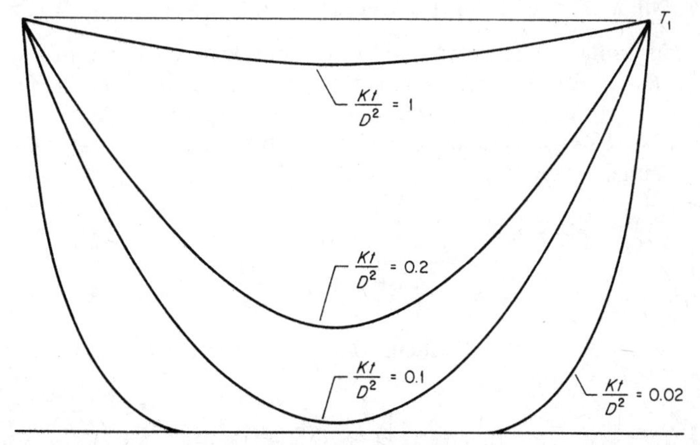

Figure 8-10. Temperature diffusion along a rod initially at temperature zero whose ends are brought to T_1 and held there after time zero, shown as a function of time.

temperature changes as a function of time in a bar. The t/D^2 term in the exponents is characteristic of all diffusion processes in which the diffusing substance is not consumed in the volume. It is not dependent upon the particular geometry of the problem. Thus, if one doubles the linear dimensions of the problem, one quadruples the time required for the temperature to reach some fixed fraction of its final value. For further references on diffusion see (5–7).

Problems

1. Derive Eq. (8.18).
2. Derive Eq. (8.25).
3. Derive Eq. (8.26).
4. Verify Eq. (8.27).
5. Solve the diffusion equation for a finite section with internal generation for the two cases $L = 0$ and $I = 0$ (see Section 9).
6. In Section 2 of this chapter we noted that a Fourier series may not be valid at the end points of the interval. Does the series (8.9) fit the function

$$f(x) = \frac{x}{H}$$

   at $x = H$ and $x = -H$?
   What is the value of the series at these end points?
   We will make use of this in the next problem.
7. If x in Problem 6 is replaced by r, the radius from the center of a sphere, and H is replaced by r_0, the quantity $r_0 f(r)/r$ is equal to one from r equal to zero to r almost equal to r_0. At r_0 it becomes zero. Use this to verify that Eq. (8.43) is a solution to (8.42) and satisfies the appropriate boundary and initial conditions.
8. If r is very small, then

$$\lim_{r \to 0} \frac{1}{r} \sin \frac{n\pi r}{r_0} = \frac{n\pi}{r_0}$$

   Using this, show that the temperature in the center of a sphere whose initial temperature is zero, is given by

$$T(0, t) = T_1 \left[1 + 2 \sum_{n=1}^{\infty} (-1)^n e^{-(n\pi/r_0)^2 Kt} \right]$$

If

$$r_0 = 10 \text{ cm}, \qquad K = 2\left(\frac{69}{\pi^2}\right)\frac{\text{cm}^2}{\text{sec}}, \qquad T(r, 0) = 0$$

how long approximately does it take the center of the sphere to reach 50° if the bulk of the sphere starts at zero temperature and the outside is held at 100°? How long does it take if K is doubled? Using the original K, how long would it take a sphere of 20-cm radius to reach 50° at its center?

9. There is an implied assumption in the derivation of Eq. (8.39). What is it?

The Theory of Blood Flow Measurement

Measurement of blood flow is of cardinal importance to both physiologists and clinicians, and a variety of ingenious methods have been developed, such as the Fick method, indicator dilution, electromagnetic flow meters, and ultrasonic flow meters. Each of these has its own advantages and disadvantages and each has often been the subject of considerable controversy regarding its accuracy.

The sources of the controversy are twofold, the accuracy of the physical measurements and the underlying mathematical assumptions inherent in each method. In this chapter we shall attempt to develop the mathematical backgrounds of the most common methods with particular emphasis on the inherent assumptions of each.

1. Time Averages and Stationary Systems

The most serious conceptual difficulty concerning blood flow measurements is that almost universally in physiological literature the mathematical basis is developed in terms of constant flow, which in many parts of the body is not even a reasonable approximation. One need hardly dwell on the fact that flow through the aortic valve is pulsatile but it is perhaps less well known that the pulsatile flow extends through the circulatory system down to the capillary level and even into the veins. In many capillary beds, at the base of the fingernail for example, flow is easily observed not only to be intermittent but to actually reverse during part of the cardiac cycle. To be at all logically rigorous, the development of the theory of blood flow measurements must take into account the pulsatile nature of the flow. In order to do this we must introduce the concept of time average, which is probably quite familiar to the reader even though he may not have used the name before. We often refer to "average" blood pressure, by which we mean the average of a large number of readings of pressure taken at intervals over a number of cardiac cycles. Similarly, we speak of "average" temperature during a day, by which we mean the average of temperatures taken at equal time intervals throughout a day. Both of these examples refer to quantities that are measured at discrete points in time. More commonly, physiological measurements are made continuously; for example, a strip chart recording is a continuous measurement. We can define a time average of a continuous function of time by breaking the interval into a very large number of equal small intervals and averaging the values obtained from each small interval. From Figure 9-1 it

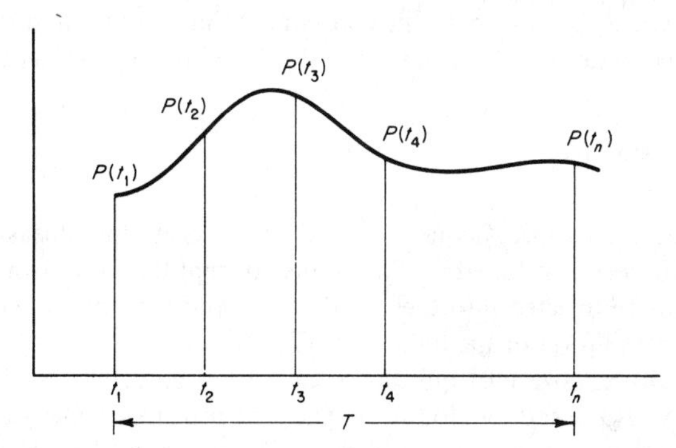

Figure 9-1. Relationship between the area under a curve and the time average.

is apparent that the average so obtained is equal to the area under the curve (measured from the zero line) divided by the total length of the time T:

$$\text{Average } P = \bar{P} = \frac{P(t_1) + P(t_2) + P(t_3) + \cdots + P(t_n)}{n}$$

$$T = t_n - t_1 = (n-1)\,\Delta t$$

For n large, $T \approx n\,\Delta t$. Then

$$\bar{P} = \frac{P(t_1) + P(t_2) + \cdots + P(t_n)}{T/\Delta t} = \frac{P(t_1)\,\Delta t + P(t_2)\,\Delta t + \cdots + P(t_n)\,\Delta t}{T}$$

Since the area under a curve between t_1 and t_n is the integral of the function that defines the height of the curve, in this case $P(t)$, the sum above can be written as

$$\bar{P} = \frac{1}{T}\int_{t_1}^{t_n} P(t)\,dt \tag{9.1}$$

Thus, the time average is

$$\bar{P} = \frac{1}{t_n - t_1} \int_{t_1}^{t_n} P(t)\, dt \tag{9.2}$$

There are many kinds of measured quantities, in particular those concerning blood flow, whose instantaneous values change rapidly as a function of time but whose time averages over sufficiently long intervals are relatively stable. Such quantities are referred to as stationary. Thus, aortic flow measured instantaneously is not stationary, but measured over a period of, say, 10 sec, is approximately stationary. Similarly, one's caloric consumption varies a great deal from hour to hour but is approximately stationary over an interval of a week.

Armed with these new tools, the concepts of time average and stationarity, we can begin to develop rigorously the theory of blood flow measurements.

2. Indicator Dilution

The technique most frequently used for blood flow measurement is indicator dilution. It is based on the simple concept that a known amount of indicator can be injected into the bloodstream and the flow rate determined by the rate of dilution of the indicator (17, 33).

Historically a variety of indicators have been used. Evans blue, cardiogreen dye, diodrast, sodium, iodine, oxygen, nitrous oxide, acetylene, carbon monoxide, radioactive krypton, xenon, and probably others that have not come to the author's attention. The choice among these is usually based on physiological properties of the indicator, the ease with which it can be detected in the bloodstream and whether or not it is subject to error due to recirculation. Dyes are easy to detect but are not quickly removed from the blood stream and recirculate, which is often troublesome. The radioactive gases, on the other hand, are easy to detect and have the advantage that they are rapidly cleared by the lungs, but have the difficulty that they are also diffusable into tissue and therefore lost from the bloodstream or stored temporarily which also causes difficulty in the interpretation of results.

Two different techniques of indicator dilution measurements are in use, the constant injection method and the impulse injection method. In the former, an indicator is injected into the bloodstream at a constant rate for a time sufficiently long so that the concentration as measured at some downstream point comes to an approximately constant value. In the latter method, impulse injection, a very brief injection is made and the time course of indicator concentration is followed at the downstream point. Both indicator dilution techniques are based on the idea of the flow concentration integral.

If the flow past a point in a system is in one direction only, the quantity of an indicator substance which passes the point during the time interval t_1 to t_2 is given by

$$Q_{t_1,t_2} = \int_{t_1}^{t_2} F(t)C(t)\,dt \tag{9.3}$$

If the flow is constant, this integral becomes

$$Q_{t_1,t_2} = F_0 \int_{t_1}^{t_2} C(t)\,dt \tag{9.4}$$

If the flow is not constant but stationary over the interval and the concentration is constant,

$$Q_{t_1,t_2} = C_0 \int_{t_1}^{t_2} F(t)\,dt = C_0(t_2 - t_1)\bar{F} \tag{9.5}$$

3. Constant Injection

If an indicator is injected into the system at a constant rate R beginning at time zero and for a long enough time so that the concentration at a downstream point becomes essentially constant, it is clear that in the case of *constant flow* the amount of indicator which crosses the measuring point in the interval t_1 to t_2 must be the same as that entering the system in the same interval of time

$$(t_2 - t_1)R = F_0 \int_{t_1}^{t_2} C(t)\,dt$$

from which we obtain

$$F_0 = \frac{R}{[1/(t_2 - t_1)] \int_{t_1}^{t_2} C(t)\,dt} = \frac{R}{\bar{C}} \tag{9.6}$$

If the flow is not constant, but is at least stationary, and the concentration is constant,

$$(t_2 - t_1)R = C_0(t_2 - t_1)\bar{F}, \qquad \bar{F} = R/C_0 \tag{9.7}$$

If neither the flow nor the concentration is constant, but both are stationary, and the rate of input of the indicator is constant, the concentration change is caused by the flow change. In order to make this point clear consider a limiting case in which the flow changes periodically between F_1 and F_2 as shown in Figure 9-2. Note that over intervals long compared to $T_1 + T_2$ the flow is stationary.

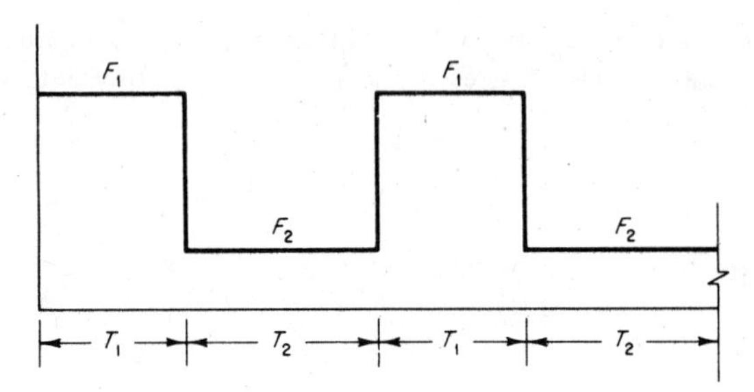

Figure 9-2. Alternate high and low flow rates.

If one measures at a point immediately downstream from the injection point before any mixing in the direction of flow has occured, one will measure due to the input which occurred during T_1 a concentration C_1 where

$$C_1 = R/F_1$$

Similarly, due to the injection which occurs in the interval T_2 one measures C_2 given by

$$C_2 = R/F_2$$

One can determine the correct average flow by adding the effects observed during the two intervals:

$$\bar{F} = \frac{F_1 T_1 + F_2 T_2}{T_1 + T_2} = R \frac{(T_1/C_1) + (T_2/C_2)}{T_1 + T_2}$$

If the flow is changing continuously instead of in steps as above, we break the time interval into n subintervals and the continuous flow into a large number of steps and have

$$\bar{F} = R \frac{\dfrac{\Delta t_1}{C_1} + \dfrac{\Delta t_2}{C_2} + \cdots + \dfrac{\Delta t_n}{C_n}}{\Delta t_1 + \Delta t_2 + \cdots + \Delta t_n}$$

which we can now convert back to the continuous form by replacing the sum by an integral

$$\bar{F} = \frac{R}{t_2 - t_1} \int_{t_1}^{t_2} \frac{1}{C(t)}\, dt \tag{9.8}$$

Note that this latter average

$$\frac{R}{t_2 - t_1} \int_{t_1}^{t_2} \frac{1}{C(t)}\, dt$$

is not the same as

$$\frac{R}{t_2 - t_1} \frac{1}{\int_{t_1}^{t_2} C(t)\, dt}$$

but if C does not change very much, it is close to being the same. Note also that this form reduces to the previous cases (9.6) and (9.7) without requiring that either flow or concentration be constant, but only that the system be stationary and the change in concentration be due to change in flow. See Problem 1 at the end of this chapter.

The constant injection method outlined above has a number of advantages. The result is dependent only upon the rate of input of indicator and the *concentration* of the effluent. Thus, it is not necessary to monitor all outputs of the system but any one can be used as a sample provided that mixing occurs before any splitting of the path. The method used with Eq. (9.8) yields the correct answer in the presence of pulsatile flow and the result is not affected by temporary storage of the indicator by tissue provided that a steady state is maintained long enough for any diffusion process to come into equilibrium. Unfortunately, it suffers from one very serious disadvantage, which is that recirculation of the indicator often makes it impossible to maintain a steady state, unless indicators which are lost on one pass through the system are used, but these tend to diffuse into tissue and have very long equilibrium times. Therefore, in spite of the advantages of this technique, impulse injection is usually preferred.

4. Impulse Injection

The impulse injection method differs from the constant rate method in that a single fast injection is used. At a point downstream from the point of injection the concentration is usually seen to rise quickly and decay relatively slowly, but both of these processes are usually slow compared to the interval between heart beats. Since all of the injected indicator must come out the flow concentration integral evaluated from the time of injection to infinity must equal the total injected indicator:

$$Q = \int_0^\infty F(t)C(t)\, dt$$

If the flow is constant

$$Q = F_0 \int_0^\infty C(t)\, dt$$

and F_0 can be found from

$$F_0 = \frac{Q}{\int_0^\infty C(t)\,dt}$$

If the flow is not constant but periodic and stationary the integral can be evaluated by breaking it up into intervals of one heart period each:

$$\int_0^\infty F(t)C(t)\,dt = \int_0^T + \int_T^{2T} + \int_{2T}^{3T} F(t)C(t)\,dt \cdots$$

If within each interval the concentration is approximately constant

$$\int_0^\infty F(t)C(t)\,dt \approx C_1 \int_0^T F(t)\,dt + C_2 \int_T^{2T} F(t)\,dt \cdots C_n \int_{(n-1)T}^{nT} F(t)\,dt$$

where C_k is average concentration in the kth interval. If the flow is stationary, each integral in the foregoing is the average flow multiplied by T. Thus,

$$Q = \bar{F}T(C_1 + C_2 + \cdots + C_n)$$

$$\bar{F} = \frac{Q}{T(C_1 + C_2 + \cdots)}$$

and this is approximately

$$\bar{F} = \frac{Q}{\int_0^\infty C_s(t)\,dt}$$

where C_s is a smooth function that approximates the series C_k.

The integral in the foregoing is indicated as being an infinite integral. In practice, the length of time over which C can be observed is limited by recirculation time. The assumption is usually made that the descending part of the curve approaches a single exponential decay and the measurement is terminated at some time just before recirculation becomes evident. The concentration is then plotted on semilogarithmic paper and the concentration extrapolated as if no recirculation had occurred. The validity of this is somewhat questionable, but it is probably the best that one can easily do. (See Problem 2 at the end of this chapter.)

5. Vascular Volume

In addition to the measurement of blood flow, indicator dilution provides a measure of the volume of a vascular bed. The logic of this determination is slightly subtle and is perhaps best illustrated by an analogy to a dance floor in a nightclub. In estimating the space required for a dance floor, one need not provide space for all the couples who might be in the nightclub at one time. Rather one need only provide space for the fraction of the couples who will be dancing at any one time. This fraction is approximately equal to the length of time each couple chooses to dance divided by the total time they spend in the nightclub, and for simplicity, let us assume that all couples stay all evening, a time that we call T_E. Thus, if each couple requires an area A in which to dance, the total area required will be

$$A_T = A\left(\frac{T_1}{T_E} + \frac{T_2}{T_E} + \frac{T_3}{T_E} + \cdots\right)$$

where the subscripts 1, 2, 3, and so on refer to the length of time the first, second, and third couple choose to dance. Now, let us return to a description of a fluid and assume that we could follow the path of an individual molecule. Associated with each molecule is a volume, which is its fraction of the total volume; also associated with each molecule is a transit time T, which is the time interval between the time it enters the system and the time it leaves the system. We define the function $P(T)$, which is the fraction of the total number of molecules in the system that have a transit time T. Thus, if all the molecules remained in the system exactly the same time T_1, $P(T_1)$ would be 1 and P of any other time would be zero. If two-thirds of the molecules have a transit time T_1 and one third a transit time T_2, then $P(T_1)$ would be two thirds, $P(T_2)$ one third, and P of any other time zero.

The number of molecules entering the system per unit time is equal to the flow divided by the average volume per molecule

$$N = \frac{\bar{F}}{v} = \frac{\text{liters/sec}}{\text{liters/molecule}} = \frac{\text{molecules}}{\text{sec}}$$

Each molecule contributes to the volume of the system for a time equal to its transit time. Therefore, in the two thirds–one third case, the volume of the system would be

$$V = v(\tfrac{2}{3}NT_1 + \tfrac{1}{3}NT_2) = F(\tfrac{2}{3}T_1 + \tfrac{1}{3}T_2)$$

Now, let us replace our simple approximation of discrete transit times T_1, T_2 by a continuous distribution of transit times $P(T)$, which we can depict roughly as in Figure 9-3. In this graph the *fraction* of molecules having a transit time between T and $T + \Delta T$ is the height of the curved represented by the shaded rectangle. The total area in the curve is the sum of all possible rectangles, and is therefore equal to the total area of the curve, which must be equal to 1. As in the discrete case, we add up the contributions to volume from molecules having transit times corresponding to each small rectangle

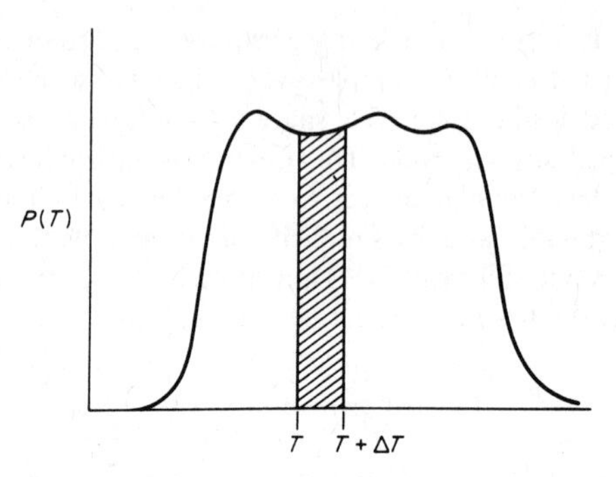

Figure 9-3. Distribution of transit times versus time.

multiplied by the transit time associated with that rectangle, and from our previous discussion of the concept of an integral we find that the total volume is given by

$$V = \bar{F} \int_0^\infty TP(T)\, dT$$

Now our problem is reduced to one of finding $P(T)$, the fraction of molecules that have each transit time. This may be done by means of an impulse injection of indicator at the input to the system and observation of the rate at which the indicator leaves the system $C(T)F$. The *fraction* of the injected indicator that leaves between time T and $T + \Delta T$ is equal to the value of $P(T)$ at T:

$$P(T) = \frac{C(T)\bar{F}}{\int_0^\infty C(T)\bar{F}\, dT} = \frac{C(T)}{\int_0^\infty C(T)\, dT}$$

Thus

$$V = \frac{\bar{F} \int_0^\infty TC(T)\, dT}{\int_0^\infty C(T)\, dT}$$

6. Parallel Compartments

The method discussed in the foregoing yields information about total flow and total volume from the integrals of the concentration versus time curve. There is another variety of information, which is sometimes obtained from the *shape* of the concentration versus time curve. The reader will recall from earlier chapters that the concentration time curve of a single-compartment dilution is a simple decaying exponential whose decay rate is given by the quantity $-F/V$. In many cases when an indicator substance is injected into an organ in a short fast bolus, the quantity of remaining indicator as a function of time is observed to differ from a simple decaying exponential. When plotted logarithmically, which yields a single straight line for simple decay, the curve consists of a very long straight tail but is found to start higher and have a much faster initial decay than would correspond to the time constant of the long tail. This is interpreted as representing parallel compartment flow of two or more compartments. Consider the model in Figure 9-4. If a bolus of indicator is injected into the common input of the

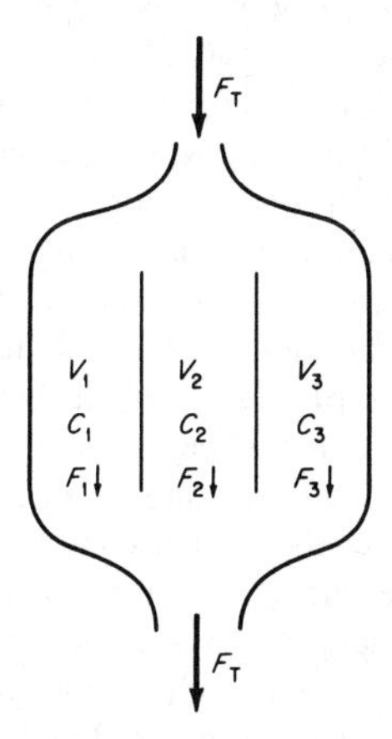

Figure 9-4. Parallel flow compartments.

parallel compartments, though each compartment may have the characteristic of a simple decaying exponential, the total amount of indicator remaining in the three compartments at time t is given by

$$Q_T(t) = Q_1(0)e^{-(F_1/V_1)t} + Q_2(0)e^{-(F_2/V_2)t} + Q_3(0)e^{-(F_3/V_3)t}$$

If we assume thorough mixing in the input, the amount of indicator initially in each compartment is proportional to the fractional flow into that compartment:

$$Q_1(0) = Q_T(0)\frac{F_1}{F_T}, \qquad F_T = F_1 + F_2 + F_3$$

$$Q_2(0) = Q_T(0)\frac{F_2}{F_T}, \qquad Q_T = Q_1 + Q_2 + Q_3$$

$$Q_3(0) = Q_T(0)\frac{F_3}{F_T}$$

$$Q_T(t) = Q_T(0)\left(\frac{F_1}{F_T}e^{-(F_1/V_1)t} + \frac{F_2}{F_T}e^{-(F_2/V_2)t} + \frac{F_3}{F_T}e^{-(F_3/V_3)t}\right)$$

Thus, if the remaining indicator versus time curve can be expressed as a sum of a number of simple decaying exponentials (the way in which this is done will be discussed momentarily), then the relative flow in each compartment is known and since total flow can be found by the integral method, the volumes are known. Note that this is much more information than one gets from the simple integral form previously discussed, where it was possible to determine only total flow and total volume.

This interpretation of washout curves is a subject of perennial vigorous controversy. At the risk of throwing some turpentine on the fires, let us at least try to explore carefully the assumptions inherent in the method and, having enumerated these, consider what biological problems are consistent with these assumptions. Assumption one is that the washout curve can be accurately represented as the sum of a small number of simply decaying exponentials. Two, that each of these starts at the same initial time, which is to say that the filling of the compartments is truly a parallel process and not a process such as might occur if one compartment were fed through a short branch of an artery while another compartment were fed from a long branch of the artery. Three, the assumption is made that the initial quantity of tracer in each compartment is proportional to the flow in that compartment. Four, that the flow through each compartment does not change with time. Five, that the entire organ is monitored and that the indicator detection process applies with equal efficiency to all the compartments.

Within these assumptions, when the curve can truly be constructed out of a moderate number, two or three or four at the most, of simply decaying exponentials, there is very little doubt in this author's mind that the compartmentalization thus determined is valid. It may not always be possible to physically localize the compartments, and in fact, they may not be compartments in the sense of parallel buckets but could, for example, be the result of chemical processes occurring at different rates in the same physical space. From the point of view of analysis, however, these considerations do not negate the validity of the method.

Much of the controversy over this method fails to distinguish between the mathematical structure of the method, which is valid, and the question of whether or not the foregoing assumptions are true. Let us take these assumptions in reverse order, leaving the most difficult until the end. A true measure of the total remaining indicator is usually not very difficult to obtain when the outflow consists of a single vein or when the entire organ can be monitored by a counter. Caution must be observed in this case to be sure that the counter sees all parts of the organ equally and treats each of them with equal counting efficiency. This is not easy to do if the organ being monitored is a large one. It is often difficult to locate a counter in such a way that all parts of a large organ will be counted with equal efficiency, but to do so is absolutely essential to the validity of the method.

The assumption of simultaneous starting times in each compartment is also sometimes difficult to validate. Let us return momentarily to the one-compartment case and consider the effect of a length of artery between the point of injection of the indicator and entry to the compartment. One would obtain a simple exponential decay, but it would not begin simultaneously with the injection of the bolus; it would be delayed by the filling time due to the length of the filling artery. In the simple one-compartment case, we would have very little difficulty in interpreting what had happened. In trying to analyze a multicompartment case, however, a delay in the filling of one of the smaller compartments would almost certainly be masked by instantaneous filling of a larger compartment. This leads to an incorrect estimate of the fractional flow. In some of the literature where this technique has been used, this point has been investigated by means of radioautographs made from tissue slices that were quick-frozen within a few seconds after injection of the tracer bolus and at least some reasonable assurance has been obtained that the compartments fill essentially simultaneously on a time scale relative to their individual washout times. In some organs to which this sort of analysis has been applied, however this author believes that is a great deal of doubt as to the validity of these analyses due to the delay in filling, for example, parts of the liver.

The most difficult problem associated with this method is that of establishing whether the observed experimental curve can truly be represented as a sum of a moderate number of simple exponential decays. There are a number of techniques in use for dissecting the experimentally observed curves, and two fairly sophisticated techniques will be discussed in Chapter X. The most intuitively obvious way, *peel back*, though not the best way, has the appeal of historical priority and conceptual simplicity and is adequate to illustrate many of the aspects of multicompartment processes.

7. The Peel-Back Method

The peel-back process consists of the following. A set of data points (with background subtracted) from a possible multicompartmental exponential is plotted on semilogarithmic paper. If the process truly consists of a sum of exponentials having significantly different decay rates, the curve will eventually produce a straight line decay on semilog paper. If plotted far enough in time, the slope of the straight line will be the decay constant F/V of the most slowly decaying exponential. This line is then extrapolated back to the time origin; traditionally, this is done by laying a piece of black surgical thread over the trail end of the curve and marking the intercept point at zero time. At each data point the value from the silk thread line is subtraced from the data point and the difference is plotted on the semilogarithmic paper. If the process is a two-compartment parallel process, these differences will yield a new straight line. If the process is a three-compartment process, these difference points will not be a straight line but will have a straight tail, as did the initial data, and a second silk thread fit is made extrapolating back to zero time. The sum of the two silk thread curves at each time is then subtracted from the data points and the difference again plotted. A single straight line results, in which case the process is said to have three compartments. Prior to the existence of electronic digital computers, the mere labor of the peel-back process was probably enough to give multicompartmental analysis a bad reputation. In particular, one frequently finds that the statistical scatter of data points causes the result on one of the subtractions to be negative, and on semilogarithmic paper there is no place to plot the log of a negative number. The method further suffers from another very serious drawback, which is extreme sensitivity to the accuracy with which the long components are plotted or the accuracy with which the silk thread is laid upon them. A very slight error in the slope of the silk thread produces a large error in the difference, which in turn then produces a much larger erroneous result in the slope determined for that component and a still larger error in the plot of the next compartment back. The first of these problems, the negative numbers, is circumvented by the other more sophisticated techniques, which will be discussed later. However, the sensitivity of the method to the accuracy of the determination of the longest time constant is still a somewhat vulnerable point.

There have been many attempts to analyze the nature of this error-buildup phenomenon and the sensitivity of the method to it. Each of these methods from the purist point of view has very severe theoretical difficulties. This author has proposed a method that he believes is at least an intuitively satisfying internal checking procedure. The data points are split into two groups, odd-numbered points being Group A and even-numbered points Group B. Each of these groups, A and B is then analyzed by separating it into simple exponentials either by manual peel back, as described above, or by either of the two methods to be described later. If the results obtained thereby are in reasonable agreement, it is very difficult to argue that there is a significant error due to statistical fluctuation. This is probably not the strongest test that could be applied, but it has a sort of sweeping inclusiveness about it that is very appealing. An example of the A, B test is given in Chapter X.

It will be recalled that we have available an independent check on the validity of multicompartment exponential analysis. The integral analyses developed earlier in this chapter are independent of the detailed nature of the flow process; therefore, the same data can be analyzed by the integral process, and the total vascular volume should agree with the multicompartmental analysis.

The author feels quite certain that the preceding discussion will not satisfy the mathematical purists. However, there is physiological information to be obtained in this world and if one waits for a perfect analysis, one may wait a long time. No scientific result is ever absolutely certain. It is only arrived at as representative of a preponderance of evidence. When the assumptions of exponential analysis are verified by radioautography and the validity of the separation into individual exponentials is verified by the A, B test described above and by agreement with the total vascular volume determined by the integral methods, it is almost certain that there is significant biological information.

8. The Fick Method

The Fick method is almost universally regarded as the most reliable method of cardiac output measurement. The basis of the measurement is that the difference in oxygen concentration of blood entering the lungs and blood leaving the lungs multiplied by the flow rate is equal to the average rate of oxygen uptake by the lungs

$$\bar{R} = \bar{F}(\bar{C}_{\text{out}} - \bar{C}_{\text{in}}) \tag{9.9}$$

where $\bar{R}$ is the average rate of oxygen uptake, $\bar{F}$ the average flow rate, and $\bar{C}_{\text{in}}$ and $\bar{C}_{\text{out}}$ the average concentrations of oxygen in blood entering and leaving the pulmonary system.

Although the technique is difficult, the method comes very close to being foolproof. The difficulty stems from the need of a reliable measure of venous blood concentration, which usually must be obtained by a catheter into the right heart. By the same token, the method is free of the uncertainties of indicator dilution in that the recirculation is actually measured and corrected for. It is also insensitive to the effects of pulsatile flow. Even though flow through the pulmonary system is quite pulsatile, the concentration of oxygen in the blood leaving the lungs is quite uniform due both to mixing and the fact that the blood is nearly saturated with oxygen. Venous blood in the right ventricle is well mixed. Thus, the flow concentration integral can be evaluated in terms of stationary flow by taking the concentrations out from under the integral sign:

$$\bar{R} = \frac{1}{t_2 - t_1}\int_{t_1}^{t_2} R(t)\,dt$$

$$= \frac{1}{t_2 - t_1}\int_{t_1}^{t_2} F(t)C_{\text{out}}(t)\,dt - \frac{1}{t_2 - t_1}\int_{t_1}^{t_2} F(t)C_{\text{in}}(t)\,dt$$

$$= \frac{C_{\text{out}} - C_{\text{in}}}{t_2 - t_1}\int_{t_1}^{t_2} F(t)\,dt = (C_{\text{out}} - C_{\text{in}})\bar{F}$$

Problems

1. Consider a case in which the flow is given by

$$F(t) = \frac{F_1}{(1 + \alpha t)^2}$$

from time zero to time T and repeats thereafter. The average flow is given by

$$\bar{F} = \frac{1}{T}\int_0^T F(t)\,dt = \frac{F_1}{T}\int_0^T \frac{1}{(1 + \alpha t)^2}\,dt$$

which is evaluated by the following trick. Let $y = 1 + \alpha t$, $dy = \alpha\,dt$

$$\bar{F} = \frac{F_1}{T}\int_1^{1+\alpha T} \frac{1}{y^2}\frac{dy}{\alpha} = \frac{F_1}{T\alpha}\left(1 - \frac{1}{1 + \alpha T}\right) = \frac{F_1}{1 + \alpha T}$$

Now assume a constant rate R of injection of an indicator resulting in a concentration just downstream from the injection point of

$$C(t) = \frac{R}{F(t)} = \frac{R}{F_1}(1 + \alpha t)^2$$

Show that the time average of the concentration is given by

$$\bar{C} = R(1 + \alpha T + \tfrac{1}{3}\alpha^2 T^2)/F_1$$

and try to use this in (9.6) to find the flow. One finds

$$F_0 = \frac{R}{\bar{C}} = \frac{F_1}{1 + \alpha T + \frac{1}{3}\alpha^2 T^2}$$

which for small values of αT, small changes in the concentration, agrees closely but not exactly with $\bar{F}$ given above. For large changes in concentration, the agreement would be poor. Now show using (9.8) that

$$\bar{F} = \frac{F_1}{1 + \alpha T}$$

the correct value.

2. In an impulse injection dilution measurement, observation was terminated at time T. From time zero to T the measured area was found to be

$$A_m = \int_0^T C(t)\,dt$$

The descending part of the curve was found to have the form

$$C(t) = C_1 e^{-\alpha(t-T)}$$

where C_1 was $C(t)$ at time T. The total concentration integral area is

$$\int_0^\infty C(t)\,dt = \int_0^T C(t)\,dt + \int_T^\infty C(t)\,dt$$

Shows that this equals

$$A_m + \frac{C_1}{\alpha}$$

3. For some years the natural gas industry has measured the rate of flow of gas through pipes by means of radioactive tracers. The technique is analogous to a dye dilution measurement of cardiac output. Let us assume that 1 gm of tracer gas is shot into a pipeline at 12 noon at South Junction. At East Liverpool, some way down the pipe, the concentration of tracer as a function of time is found to be as shown in Figure 9-5.

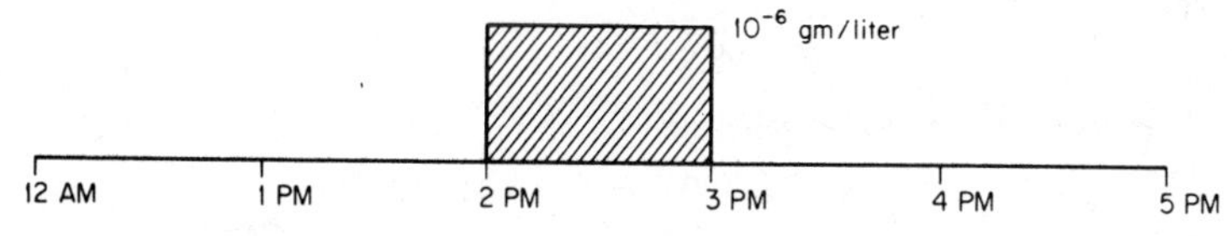

Figure 9-5. Concentration of tracer measured in a natural gas pipe at East Liverpool.

What is the rate of flow of natural gas through this pipe? What is the volume of pipe between East Liverpool and South Junction?

4. One gram of tracer is introduced into a flow system at time zero. The concentration in the ouput is observed to be

$$C(t) = t^2 e^{-\alpha t} 0.2 \frac{\text{gm}}{\text{cm}^3 \text{ sec}^2}$$

$$\alpha = 0.1 \text{ sec}^{-1}$$

What is the flow and the volume of the system? Hint:

$$\int_0^\infty t^2 e^{-\alpha t}\, dt = \frac{2}{\alpha^3}, \qquad \int_0^\infty t^3 e^{-\alpha t}\, dt = \frac{6}{\alpha^4}$$

5. The following points were taken from a two-component decay curve. Try to find the two components by means of the peel-back process.

Time (sec)	Value
0	110
1	59
2	33
3	20
4	12.8
5	9.0
6	6.8
7	5.4
8	4.6
9	3.9

6. To the data in Problem 5 add the following and try a peel-back separation.

Time (sec)	Value
10	3.4
11	3.0
12	2.7
13	2.4
14	2.1

Answers: Using the Prony method described in Chapter X, the first 10 points yielded

$$Y = 98.9e^{-0.70t} + 11.0e^{-0.12t}$$

The data were actually generated from

$$Y = 100e^{-0.69t} + 10.0e^{-0.10t}$$

What does this tell you about the accuracy of the peel-back process?

Curve Fitting

1. Partial Derivatives

Imagine you are standing on the side of a mountain—a ski slope, for example—and you wish to characterize the slope of the mountain at the particular point at which you are standing. Having read the first chapter of this book, you know that one describes the slope of a curve by the derivative at each point along the curve, and that the derivative is defined by taking a small step in the independent variable and finding how much the dependent variable changes as a result of the small step. On the side of the mountain, however, the problem is a little more complicated because it may make a difference in what direction the small step is taken. In this case there are two independent variables: one that is a measure of east–west position, which we shall call x, and the other, at a right angle, a measure of north-south position, which we shall call y. If the slope of the mountain runs from south to north, a step along y in the positive direction will result in a negative step in height. On the other hand, a step in the x direction will result only in a small change of height. We characterize this by *partial derivatives. The partial derivative with respect to y is a measure of the change in the height of the mountain if we take a step in the y direction, keeping the x coordinate constant. The partial derivative with respect to x is measured by taking a step in the x direction, keeping the y coordinate constant.*

It is not obvious that these two values, the partial derivative with respect to x and the partial derivative with respect to y, fully characterize the slope at a point. However, by analogy to fitting a tangent line to a curve, as we did in Chapter I, we can fit a tangent plane to the side of a mountain at one point. Imagine, for example, a flat sheet of plywood. The angle of the sheet of plywood is fully described by its slope in two perpendicular directions. If this is difficult to visualize, take a flat sheet of cardboard, such as a tablet back that has square corners, and prop it up on a desk in such a way that you have a one-inch slope along one side and a two-inch slope along the perpendicular side. You will notice that these two values fully describe the slope of the sheet of cardboard.

Now let us formally define partial derivatives. Let f be a function of x and y where f, for example, could stand for the height of a mountain and the x and y coordinates are in the east and north directions. *The partial derivative with respect to x is defined as the change in f that occurs due to an infinitesimal step in x, keeping y constant, divided by the size of the step.* To distinguish partial derivatives from total derivatives, a somewhat different symbolism is used. The letter d, which was used as the symbol for a total derivative, is replaced by the symbol ∂, an alternative form of the greek delta δ.

$$\frac{\partial f}{\partial x} = \lim_{\substack{\partial x \to 0 \\ y \text{ constant}}} \frac{f(x + \partial x, y) - f(x, y)}{\partial x}$$

Similarly, *the partial derivative with respect to y is the change that occurs in f due to an infinitesimal step in y, keeping x constant, divided by the size of the y step.*

$$\frac{\partial f}{\partial y} = \lim_{\substack{\partial y \to 0 \\ x \text{ constant}}} \frac{f(x, y + \partial y) - f(x, y)}{\partial y}$$

When one computes a partial derivative with respect to x, all other variables, y, for example, are treated as if they are numerical constants. Thus, to compute the partial derivative with respect to x of the quantity

$$f(x, y) = x^2 + y^2 + 3xy$$

we treat y as if it were a number

$$\frac{\partial f}{\partial x} = 2x + 3y$$

The partial derivative with respect to y treats x as a constant

$$\frac{\partial f}{\partial y} = 2y + 3x$$

2. Maximization and Minimization

The reader will no doubt remember from elementary calculus that if $y = f(x)$, maximum values of y and minimum values of y are determined by finding those values of x for which the derivative of y with respect to x is equal to zero. Perhaps it is well to review why this is so. Let $f(x)$ be any continuous function of x that has a derivative at every point between x_1 and x_2. By the phrase "has a derivative" we mean that the function has no sharp corners, so that its slope can be determined at every point. The condition for a local maximum or a local minimum is that y, as a function of x, must reverse direction at this value of x. In the process of reversing direction, it must go through a point at which its slope is zero. (If necessary, the reader may want to review Chapter I.) Such points are not, in an absolute sense, maxima or minima. They are what are called local maxima and minima, because it is quite possible that the function reverses directions many times, as the function shown in Figure 10-1, in which we find both a local maximum at $x = -2$

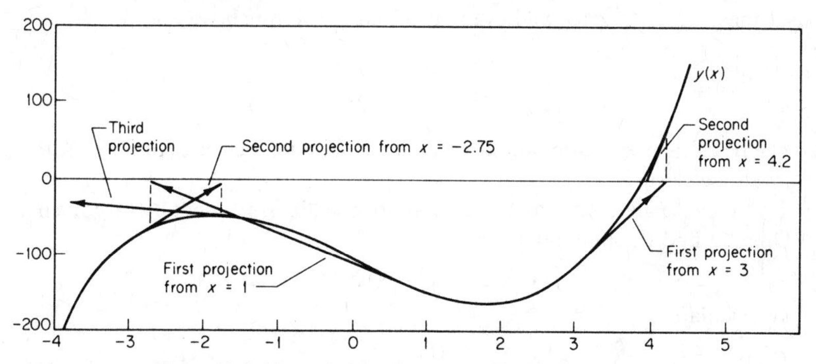

Figure 10-1. Numerical solution of an algebraic equation.

and a local minimum at $x = 2$, neither of which is, in an absolute sense, a maximum or minimum.

The problem with which we are concerned in this chapter involves finding minimum values of a function of more than one variable. Let us define the function $z = f(x, y)$, and let us establish the condition for a local minimum of this function. In order to do so, let us assume that the values of x and y at which a local minimum occurs are known and are, respectively, a and b. If the point a, b is a local minimum, then certainly the function $f(a, y)$ (a fixed, y variable) has a minimum when $y = b$, at which point the derivative of f with respect to y is equal to zero. Now, let us reverse the argument, and evaluate $f(x, b)$ (x variable, b fixed). By the same logic, where $f(x, b)$ has a local minimum, the derivative of f with respect to x is equal to zero. For the function $f(x, y)$ to have a local minimum, both conditions must be simultaneously satisfied, and these derivatives are identical to the partial derivatives. We have, therefore, the conditions

$$\frac{\partial f(x, y)}{\partial x} = 0, \qquad \frac{\partial f(x, y)}{\partial y} = 0$$

where once again we state that a partial derivative with respect to a specific variable is calculated in the same manner as a conventional derivative, except that all other variables are treated as numerical constants. The following example will perhaps make this clear.

Example. Consider the function

$$z = x^2 + 3x + xy + y^2$$

To find its local minimum, we calculate the two partial derivatives,

$$\frac{\partial z}{\partial x} = 2x + 3 + y, \qquad \frac{\partial z}{\partial y} = x + 2y$$

set these equal to zero, and evaluate x and y, for which we find

$$x = -2, \qquad y = 1$$

$$z_{\min} = 2^2 - 3 \cdot 2 - 2 + 1 = -3$$

which is clearly a minimum, rather than a maximum, because for x or y large, z becomes large.

We can extend this to three or more variables by identical arguments. Thus, to find the minimum of

$$f(x, y, z) = x^2 + xyz + z^2 + 2z$$

we calculate

$$\frac{\partial f}{\partial x} = 2x + yz, \qquad \frac{\partial f}{\partial y} = xz, \qquad \frac{\partial f}{\partial z} = xy + 2z + 2$$

and set each of these to zero, from which we find

$$z = -1, \qquad x = 0, \qquad y = 0$$

$$f_{\min} = -1$$

3. Differentials in More than One Variable

In the first chapter of this book we showed how to find an approximate value for the change in a function of one variable when the variable changes by a small amount Δx. We called this change Δf the differential. We recall that differentials were related to the derivative of the function calculated at x, and that we found this result from the definition of the derivative

$$\frac{df(x)}{dx} = \operatorname*{limit}_{\Delta x \to 0} \frac{\Delta f}{\Delta x} = \operatorname*{limit}_{\Delta x \to 0} \frac{f(x + \Delta x) - f(x)}{\Delta x}$$

$$\Delta f \approx \Delta x \frac{df(x)}{dx}$$

and demonstrated graphically why it was true by drawing a little triangle, as shown in Figure 1-5. In this chapter, we shall be concerned with how functions of more than one variable change for small changes in the variables. Let us consider the function f of two variables A and B. Temporarily we may regard B as fixed, and find that for a small change in A, f will change by the amount

$$\Delta f_A(A, B) = \Delta A \frac{\partial f(A, B)}{\partial A} = \Delta A \lim_{\Delta A \to 0} \frac{f(A + \Delta A, B) - f(A, B)}{\Delta A}$$

where we have subsripted the Δf by A to indicate that this is the change produced by changing A, holding B constant. Alternatively, we could have held A constant and computed a change Δf_B due to changing B. Now one might reasonably guess that the total effect of changing both A and B would be the sum of these two changes,

$$\Delta f = \Delta A \frac{\partial f(A, B)}{\partial A} + \Delta B \frac{\partial f(A, B)}{\partial B}$$

This is not quite right because if A is changed first, when B is changed, we should have calculated the partial derivative with respect to B, not at the values A, B, but at $A + \Delta A$ and B.

$$\Delta f = \Delta A \frac{\partial f(A, B)}{\partial A} + \Delta B \frac{\partial f(A + \Delta A, B)}{\partial B}$$

However, if ΔA is sufficiently small, and if the function is sufficiently smooth, then the value of the derivative calculated at $A + \Delta A$ will not be significantly different from the value calculated at A, and therefore our guess above is essentially correct.

Had we changed ΔB first, we would have had an entirely analogous argument that the exact value of the change would be

$$\Delta f = \Delta B \frac{\partial f(A, B)}{\partial B} + \Delta A \frac{\partial f(A, B + \Delta B)}{\partial A}$$

but once again, if the function is sufficiently smooth, this will not be significantly different from

$$\Delta f = \Delta A \frac{\partial f(A, B)}{\partial A} + \Delta B \frac{\partial f(A, B)}{\partial B}$$

What do we mean by "significantly different"? We remember that the process of calculating the differential in f for small changes in A or B is only an approximate process, as illustrated in Figure 1-5. It is the approximation of a straight line to a curve, and as long as the curve is sufficiently smooth, it will not be a bad approximation. By the same token, the change in the derivatives with respect to A due to a change in B will usually be exceedingly small; therefore, the result is a valid approximation. This result can easily be extended to an arbitrary number of variables, which we now call A_K, where K may take on values 1, 2, 3, ..., N, instead of A and B. The differential of the function f is given by

$$\Delta f = \sum_{K=1}^{N} \Delta A_K \frac{\partial f}{\partial A_K} \tag{10.1}$$

where we have introduced, once again, the summation notation for

$$\Delta f = \Delta A_1 \frac{\partial f}{\partial A_1} + \Delta A_2 \frac{\partial f}{\partial A_2} + \cdots + \Delta A_N \frac{\partial f}{\partial A_N}$$

4. The Solution of Nonlinear Equations

In this chapter, we shall have to solve nonlinear systems of equations. The ones we shall encounter will involve more than one unknown. For the purpose of illustration, however, let us solve a nonlinear equation in one unknown, keeping in mind that our purpose is to solve systems of multiple unknowns.

Let us consider the relatively innocuous-looking algebraic equation

$$0.01x^5 + 5.0x^3 - 50x - 100 = 0$$

This is an equation in only one unknown. However, it happens to be of a type for which there is no systematic solution. The obvious way to get an approximate solution is to plot a graph of

$$y = 0.01x^5 + 5.0x^3 - 50x - 100$$

which we have done in Figure 10-1, and we instantly observe that there is a solution somewhere near $x = 3.9$. Since we are going to have to solve for multiple unknowns, where we cannot plot graphs easily, let us see how we might have solved this numerically.

Let us assume that we know that there is a solution somewhere near the value $x = 3$, where $y = -112$. If we could draw a tangent line to the curve around $x = 3$ and project this tangent line up to $y = 0$, we might expect that the resulting change in x along this projection should bring us closer to the exact solution. Let us look at this geometrically (Figure 10-2). Suppose we compute the value of y for $x = 3$ and $x = 3.1$, for which we find the values of y: -112.5 and -103.2. We then do a simple geometrical construction, projecting up to $y = 0$. The horizontal side of the triangle is found to be of length

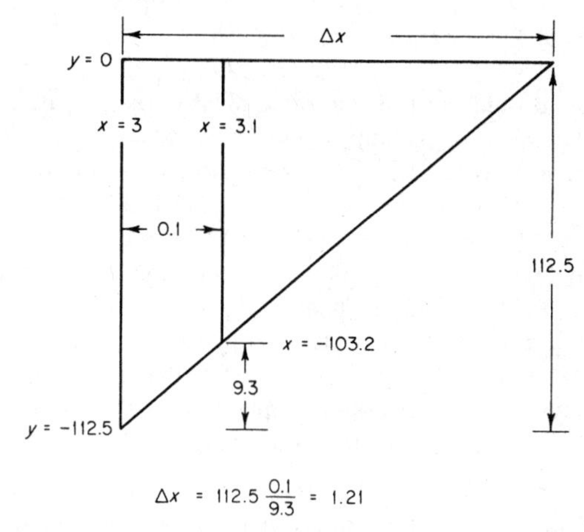

Figure 10-2. Extrapolation of trial solution.

1.21. We therefore try, as a new value for x, the value 4.21, for which we find the value of y to be 75.9. We draw a second triangle, using values of x: 4.1 and 4.2, and project backward toward $y = 0$. We find that this process rapidly converges to the root of the equation, 3.80. Now in this case, we were able to plot a graph and from this determine that $x = 3$ was a good value to start from. Suppose we had tried this process but had made the unfortunate initial guess of $x = +1$. We would find that our first projection would take us back to $x = -2.7$. Trying this value, we repeat the process and find that the next projection takes us to $x = -1.65$. We do it again and we find a value of about -5, and the process keeps going around in a circle, never getting us any closer to the correct root, around 3.8. The algebra expert would instantly say at this point that we have failed to take advantage of an obvious characteristic of this function, which is that for large values of x, it must become positive, and that therefore we know that it has to have gone through zero somewhere between $x = 2$ and x equal to a very large number, and that we were foolish to start at $x = 1$. Certainly, in a one-variable case like this, the appropriate starting points are usually obvious. In cases involving multiple unknowns, they are nowhere near so obvious, and in fact, what made them so obvious in this case was that we happen to have a particularly simple form for y. Yes, a polynomial like this is regarded mathematically as a very simple function. In spite of this, it has illustrated the method of solution, which is to calculate values of y for various values of x and assume that a straight-line projection toward the desired value of y will yield an improved value of x. It has also illustrated the possible difficulties that one can encounter. The process is as easy as rolling down a hill, provided that you happen to start on the right hill.

5. The Method of Least Squares

Unlike most problems, in which the difficulty lies in finding one answer, the problem of curve fitting is often to avoid finding too many answers. Given a set of experimental points, which for generality we shall simply identify as $Y(X_J)$, where X_J is a sequence of independent variable values X_1, X_2, X_3, ..., X_N and Y, the measured dependent variable, our problem is to find one of the many mathematical functions that will approximately fit the experimental points. Though it is not obvious, there is in reality an infinite variety of functions that will fit any set of experimental data points. In this embarrassment of plenty, what is usually wanted is a function that is in some way unique. Uniqueness may be a uniqueness of simplicity or a uniqueness due to some underlying known physical structure of the problem. An example will perhaps make this clear.

Example. Suppose one has observed a tracer experiment yielding the usual concentration as a function of time. One might choose to fit such a curve with the function

$$y = \frac{A}{1 + \sqrt{t/T}}$$

where A is the time zero value and T the interval between measurements. From t zero to $t = 10T$, this curve is almost indistinguishable from

$$y = A(0.6e^{-0.72t/T} + 0.4e^{-0.049t/T})$$

We would almost certainly choose the latter, which we can interpret as two parallel compartments, rather than the former, for which an explanation would be quite hard to find.

Given a set of data points and a functional relation that we believe will fit these data points, our problem is to adjust the available parameters of the function to get, in some sense, the best fit to the data points. Let us illustrate with a concrete example. Suppose Y equals $A + BX$ and we have a set of data points that approximate this relationship.

$$X_1 = 2, \qquad Y_1 = 5$$

$$X_2 = 4, \qquad Y_2 = 6.1$$

$$X_3 = 6, \qquad Y_3 = 6.9$$

$$X_4 = 8, \qquad Y_4 = 8$$

One way of finding appropriate values for A and B would be to plot the data points on rectangular graph paper and to lay a black silk thread on the paper in such a way as to draw what we consider to be the best straight line through the data points. Such a line is in fact often a very good approximation. However, this technique has the disadvatage that somebody else might draw a somewhat different line for what he considers to be the best fit. We are in need of some objective criterion of fit, and the one that is almost invariably chosen is the criterion of least-squared error, which is as follows. For each possible straight line, we calculate the sum of the squares of the vertical distance between the line and the data points. We then choose the line that minimizes the sum of squared errors. We do not actually have to draw all possible lines; in fact, we have to draw none of the lines, as the method of least squares uniquely defines the best possible line in this sense. The least-squared error criterion, as it is called, has certain unique properties that cause it to be chosen over other possible criteria. For example, one might choose the least sum of absolute values of errors, but there are sound theoretical reasons for preferring least-squares, in addition to which it yields mathematical forms that are considerably easier to manipulate than do other criteria.

Returning to our example, the technique of finding the best fit line is to express in terms of our parameters A and B the error that would occur at each data point for specific values of A and B. Thus, the error at data point one is

$$E_1 = Y_1 - (A + BX_1)$$

where Y_1 is the observed value of Y at X_1 and A and B are to be adjusted. E_1 is usually called the *residual.* The residual at data point two is

$$E_2 = Y_2 - (A + BX_2)$$

and so on. Each residual is then squared, so that

$$E_1{}^2 = [Y_1 - (A + BX_1)]^2$$
$$E_2{}^2 = [Y_2 - (A + BX_2)]^2$$
$$\vdots$$

and the sum of these squared errors is constructed as follows

$$S = \sum_{J=1}^{N} [Y_J - (A + BX_J)]^2$$

If we had perfect data, data that exactly fitted a straight line, the residuals would all be zero for the correct choice of A and B. We could find A and B by solving any two of the residual equations and then (for perfect data) all of the residuals would be zero. With imperfect data, the problem is that we have too many residuals. If we set any two of them equal to zero, in general, we shall find that the others are not zero. The method of least squares will give us two equations in two unknowns A and B and their solution will be the closest approximation that we can get to having all the residuals equal to zero, at least in the sense of the least-squares sum.

To minimize the sum S, we proceed by the methods of differential calculus. Notice that there are two parameters to be adjusted, A and B. If there were one parameter A, we would know that to find the value of A that minimizes the value of the sum S, we would differentiate S with respect to A, set the derivative equal to zero, and solve for A. We do an exactly analogous process in the case of two parameters, except that we now find two different derivatives, called partial derivatives, one of which is a derivative with respect to A in which we treat B as if it were a fixed numerical constant, and the other a partial derivative with respect to B in which we treat A as a fixed numerical constant. Both of these derivatives are then set equal to zero and the resulting equations are solved for A and B. The foregoing rule of handling minimizations (or maximizations) of functions of more than one variable is a general rule. The rule says to construct the partial derivatives and to set each of these equal to zero as is done in Section 1.

To find the partial derivative of the sum S, let us write out the terms

$$S = [Y_1 - (A + BX_1)]^2 + [Y_2 - (A + BX_2)]^2 + [Y_3 - (A + BX_3)]^2 + \cdots + [Y_N - (A + BX_N)]^2$$

Consider the third term and let it equal p^2

$$p^2 = [Y_3 - (A + BX_3)]^2$$

The derivative of p^2 with respect to A is

$$\frac{\partial p^2}{\partial A} = 2p \frac{\partial p}{\partial A}$$

where the partial derivative with respect to A is found in the usual fashion, recalling that B is treated as a numerical constant and that Y_3 and X_3 are constants, since they are observed values:

$$p = Y_3 - (A + BX_3), \qquad \frac{\partial p}{\partial A} = -1$$

Thus,

$$\frac{\partial}{\partial A}[Y_3 - (A + BX_3)]^2 = -2p = -2[Y_3 - (A + BX_3)]$$

The same logic applies to each term, so that

$$\frac{\partial S}{\partial A} = -2\{[Y_1 - (A + BX_1)] + [Y_2 - (A + BX_2)] + \cdots\}$$

$$= -2 \sum_{J=1}^{N} [Y_J - (A + BX_J)]$$

The partial derivative of S with respect to B is somewhat more complicated. We will find it without writing out all the terms of the sum, so that the reader will become familiar with the idea of handling derivatives of series, but if this becomes difficult to follow, write out the individual terms. Let

$$q_J = Y_J - (A + BX_J)$$

Then

$$S = \sum_{J=1}^{N} q_J^2$$

$$\frac{\partial S}{\partial B} = \sum_{J=1}^{N} 2q_J \frac{\partial q_J}{\partial B}$$

$$\frac{\partial q_J}{\partial B} = -X_J$$

Note that this step is different from the analogous step in finding the partial derivative with respect to A. Thus

$$\frac{\partial S}{\partial B} = \sum_{J=1}^{N} -2X_J[Y_J - (A + BX_J)]$$

The condition for minimizing S is

$$\frac{\partial S}{\partial A} = 0 = -2\sum_{J=1}^{N} [Y_J - (A + BX_J)]$$

$$\frac{\partial S}{\partial B} = 0 = -2\sum_{J=1}^{N} X_J[Y_J - (A + BX_J)]$$

To satisfy this condition, we take the sums apart so that we can write the unknowns in the familiar form of simultaneous linear equations

$$\sum_{J=1}^{N} (A + BX_J) = \sum_{J=1}^{N} Y_J$$

$$\sum_{J=1}^{N} (X_J A + BX_J^2) = \sum_{J=1}^{N} X_J Y_J$$

Note that the upper left sum consists of N terms with A included in each term.

The equations to be satisfied are

$$AN + B\sum_{J=1}^{N} X_J = \sum_{J=1}^{N} Y_J$$

$$A\sum_{J=1}^{N} X_J + B\sum_{J=1}^{N} X_J^2 = \sum_{J=1}^{N} X_J Y_J \qquad (10.2)$$

These are called the normal equations of the method of least squares. They are numerical equations that can be solved by Gauss reduction (described in Chapter IV, Section 1). The only thing unusual about them is the existence of summation terms instead of numerical constants preceding each of the unknowns A and B.

Let us solve the given example:

$$N = 4$$

$$\sum_{1}^{4} X_J = 2 + 4 + 6 + 8 = 20$$

$$\sum_{1}^{4} Y_J = 5 + 6.1 + 6.9 + 8 = 26$$

$$\sum_{1}^{4} X_J^2 = 4 + 16 + 36 + 64 = 120$$

$$\sum X_J Y_J = 2 \times 5 + 4 \times 6.1 + 6 \times 6.9 + 8 \times 8 = 139.8$$

These values are inserted into Eqs. (10.2), yielding

$$4A + 20B = 26$$

$$20A + 120B = 139.8$$

which are then solved by Gauss reduction, from which we find

$$A = 4.05, \qquad B = 0.49$$

The line

$$Y = 4.05 + 0.49X$$

is shown with the data points in Figure 10-3.

Sometimes we wish to treat some of these measurements as being more important or more reliable than others. We refer to this as *weighting* the observations. In order to do so, we modify the sum of squared errors by multiplying each term by a weighting function W_J. The remainder of the derivation of the normal equations proceeds along lines analogous to that already shown, yielding the result

$$\begin{aligned} A\sum_{J=1}^{N} W_J + B\sum_{J=1}^{N} W_J X_J &= \sum_{J=1}^{N} W_J Y_J \\ A\sum_{J=1}^{N} W_J X_J + B\sum_{J=1}^{N} W_J X_J^2 &= \sum_{J=1}^{N} W_J X_J Y_J \end{aligned} \tag{10.3}$$

The foregoing methods are simple in principle, and usually convenient in application. Unfortunately, the world does not consist of things that are fit by straight lines. Most problems require some kind of nonlinear curve fitting.

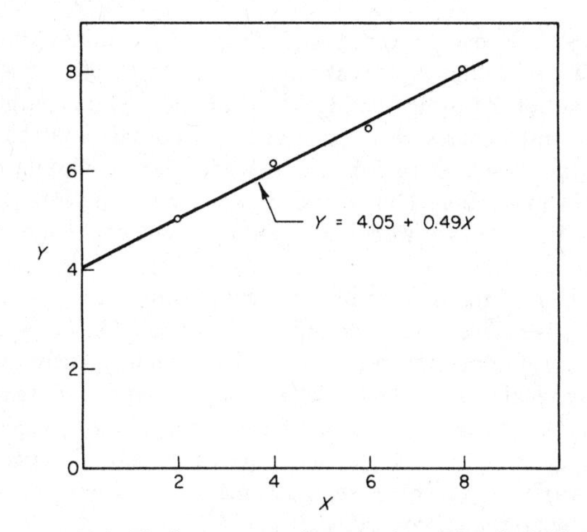

Figure 10-3. The least-squares fit to the data points in Table X-1.

6. Exponential Fitting

The most commonly encountered nonlinear fit in biology is the simple exponential decay. This is usually fitted either by plotting on semilogarithmic paper and drawing the best straight line, or by applying the method of least squares to the log of the data values, which should yield a straight line plotted against the independent variable.

Let Y_J be the logarithm of the Jth data point D_J. By means of the method of least squares, find the best straight line fit of the Y_J to the independent variable X_J. Express this in terms of the D_J by finding the antilog of the best straight line fit.

The values thus found are not exactly those which yield the minimum residuals, since the best fit for the logarithm of a function is not quite the best fit for the function. Usually no significant errors results from this approximation, unless the data scatter rather badly, in which case there is a tendency to draw the slope of the line a little bit too steep. This is evident, if the reader will consider the nature of the scales on semilog paper, which cause downward displacements of the data points to be exaggerated. There is one particular case that is occasionally troublesome. Many such curves are the result of a radioactive counting process. The statistics of radioactive counting are not those of the well-known bell-shaped curve, or Gaussian distribution. For high count rates, they approximate the Gaussian distribution closely, and therefore errors tend to be scattered approximately symmetrically above and below the true values. For low count rates, however, the statistics change to what is called a *Poisson distribution*, which is characterized by values below the mean occurring more often than values above the mean. This effect, combined with the approximation of using the logarithm of the function rather than the function, frequently causes the slope to be drawn considerably too steeply. To avoid this, a good rule of thumb is to consider as invalid data obtained at counting rates lower than about 20 counts per counting interval. In some cases, it may be necessary to combine the observed count rate of adjacent intervals to avoid this.

The problem of the multicomponent exponential decay is considerably more difficult. We have already described in Chapter IX the peel-back process, and have cautioned the reader about its accuracy. There exists a considerably better analytical method of separating multicomponent exponentials, known as the *Prony method*, which we illustrate for the two-component case.* Prony's strategy is as follows. For data spaced at equal intervals T of the independent variable (which we shall assume is time), one can replace the exponentials by powers of constants U_1 and U_2

$$\begin{aligned} e^0 &= 1, & e^0 &= 1 \\ e^{-\alpha_1 T} &= U_1, & e^{-\alpha_2 T} &= U_2 \\ e^{-\alpha_1 2T} &= U_1^2, & e^{-\alpha_2 2T} &= U_2^2 \\ &\vdots & &\vdots \\ e^{-\alpha_1 NT} &= U_1^N, & e^{-\alpha_2 NT} &= U_2^N \end{aligned}$$

We then write as the system of equations to be satisfied as closely as possible

$$\begin{aligned} Y_0 &= A_1 e^0 + A_2 e^0 = A_1 + A_2 \\ Y_1 &= A_1 e^{-\alpha_1 T} + A_2 e^{-\alpha_2 T} = A_1 U_1 + A_2 U_2 \\ Y_2 &= A_1 e^{-\alpha_1 2T} + A_2 e^{-\alpha_2 2T} = A_1 U_1^2 + A_2 U_2^2 \\ &\vdots \\ Y_N &= A_1 e^{-\alpha_1 NT} + A_2 e^{-\alpha_2 NT} = A_1 U_1^N + A_2 U_2^N \end{aligned} \tag{10.4}$$

(Note that the first data point is called Y_0, so that there are $N+1$ points. This is done to simplify the notation to follow.) From these we construct a new set of a equations by multiplying the first equation in (10.4) by 1, the second by a constant M_1, and the third by a constant M_2, and adding the three resulting equations, thus obtaining the first equation in (10.5).

We construct another new equation by multiplying the second equation by 1, the third equation by M_1, and the fourth equation by M_2 to obtain the second equation in (10.5). We construct the third new equation by multiplying the third equation in (10.4) by 1, the fourth equation by M_1, the fifth equation by M_2, and so on.

*Modern computers have made Prony's method obsolete except for starting values. The reader may read paragraph 3 on page 228 and Section 10.

$$\begin{aligned}
Y_0 + M_1 Y_1 + M_2 Y_2 &= A_1(1 + M_1 U_1 + M_2 U_1{}^2) \\
&\quad + A_2(1 + M_1 U_2 + M_2 U_2{}^2) \\
Y_1 + M_1 Y_2 + M_2 Y_3 &= A_1 U_1(1 + M_1 U_1 + M_2 U_1{}^2) \\
&\quad + A_2 U_2(1 + M_1 U_2 + M_2 U_2{}^2) \\
&\vdots \\
Y_{N-2} + M_1 Y_{N-1} + M_2 Y_N &= A_1 U_1{}^{N-2}(1 + M_1 U_1 + M_2 U_1{}^2) \\
&\quad + A_2 U_2{}^{N-2}(1 + M_1 U_2 + M_2 U_2{}^2)
\end{aligned} \tag{10.5}$$

Here is the trick: The terms within the parentheses in the Eqs. (10.5) can be set equal to zero by writing the quadratic equation.

$$1 + M_1 U + M_2 U^2 = 0 \tag{10.6}$$

where U_1 and U_2 will be the two roots of this quadratic equation. We do not know, as yet, the values of the M's, but we will find them by the method of least squares. Since the parenthesized terms in Eqs. (10.5) are zero, the left-hand side of each of the equations in (10.5) may be set equal to zero. Once more, as in the case first described in this chapter, we have too many equations to be exactly satisfied. However, we can find the best values of the M's by regarding the left sides of (10.5) as residuals to be minimized in the sense of least squares:

$$S = \sum_{J=0}^{N-2} (Y_J + M_1 Y_{J+1} + M_2 Y_{J+2})^2$$

We now calculate the partial derivatives with respect to the parameters M_1 and M_2

$$\frac{\partial S}{\partial M_1} = 2 \sum_{J=0}^{N-2} Y_{J+1}(Y_J + M_1 Y_{J+1} + M_2 Y_{J+2})$$

$$\frac{\partial S}{\partial M_2} = 2 \sum_{J=0}^{N-2} Y_{J+2}(Y_J + M_1 Y_{J+1} + M_2 Y_{J+2})$$

Set these equal to zero, from which we find the normal equations, which are

$$\begin{aligned}
M_1 \sum_{J=0}^{N-2} Y_{J+1}^2 + M_2 \sum_{J=0}^{N-2} Y_{J+1} Y_{J+2} &= -\sum_{J=0}^{N-2} Y_J Y_{J+1} \\
M_1 \sum_{J=0}^{N-2} Y_{J+1} Y_{J+2} + M_2 \sum_{J=0}^{N-2} Y_{J+2}^2 &= -\sum_{J=0}^{N-2} Y_J Y_{J+2}
\end{aligned} \tag{10.7}$$

These equations are now solved to determine the M's and the quadratic equation (10.6) for U_1 and U_2 is then solved. The values of U_1 and U_2 are now inserted into (10.4), which yields a set of simultaneous linear equations in A_1 and A_2 that must also be solved by the method of least squares

$$\begin{aligned} A_1 + A_2 &= Y_0 \\ A_1 U_1 + A_2 U_2 &= Y_1 \\ A_1 U_1^2 + A_2 U_2^2 &= Y_2 \\ &\vdots \\ A_1 U_1^N + A_2 U_2^N &= Y_N \end{aligned}$$

The normal equations are

$$\begin{aligned} A_1 \sum_{J=0}^{N} U_1^{2J} + A_2 \sum_{J=0}^{N} U_1^J U_2^J &= \sum_{J=0}^{N} U_1^J Y_J \\ A_1 \sum_{J=0}^{N} U_1^J U_2^J + A_2 \sum_{J=0}^{N} U_2^{2J} &= \sum_{J=0}^{N} U_2^J Y_J \end{aligned} \tag{10.8}$$

The foregoing technique looks very pretty. The method is readily extended to systems having more than two exponentials. For the general case of K exponentials, we have the normal equations for determining the M_I:

$$\sum_{I=1}^{K} M_I \sum_{J=0}^{N-K} Y_{J+I} Y_{J+P} = -\sum_{J=0}^{N-K} Y_J Y_{J+P}, \qquad P = 1, 2, 3, \ldots, K \tag{10.9}$$

Note that this one line represents K different equations, one for each of the values of P from 1 to K. Having solved for the M_I, we then find the values of U, which lie between 0 and 1, that satisfy an equation analogous to (10.6)

$$1 + \sum_{I=1}^{K} M_I U^I = 0 \tag{10.10}$$

This is not particularly difficult; since we know that the roots lie between 0 and 1, we can find them numerically as described in Section 1.

7. The Flaw in Prony's Method

Prony's method, as outlined above, frequently fails with real data. The difficulty is that when one fits the best line to scattered data, the data points tend to straddle the line. They do not necessarily alternate on either side of the line, but there is the moderately high probability that they will do so. According to Prony's method, all of the roots of (10.10) should lie between 0 and +1, but consider what happens if one of the roots is a negative number. Even powers of this number are positive. Odd powers are negative. The existence of a negative U causes the Prony fit to oscillate around the curve that is produced

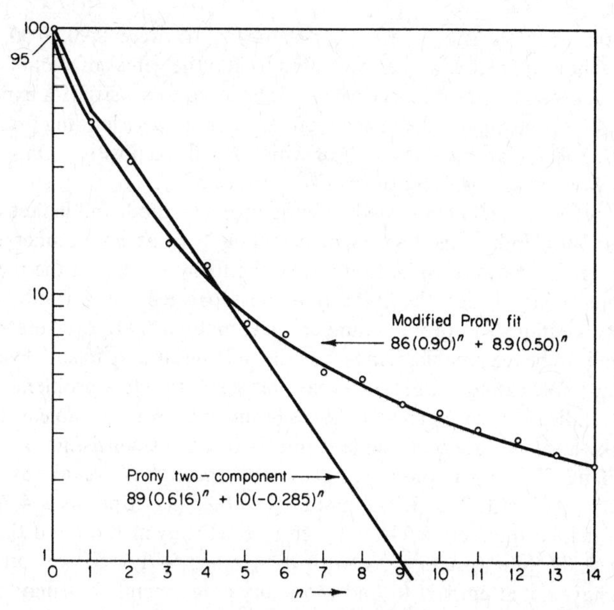

Figure 10-4. A two-component data curve separated by the Prony method looking for two components, and by the modified Prony method looking for four components of which two will be regarded as real.

by positive values of U. Thus, if one has data points such as are shown in Figure 10-4, which are, in fact the result of a two-component exponential (with scatter artificially introduced), Prony's method will fit the larger exponential with one of the values of U and oscillate around it by determining a negative value for the other value of U. The method has achieved the best least-squares fit by yielding an unsatisfactory answer. The smaller component is totally lost.

8. The Modified Prony Method

There is, however, a cure for this. It is to modify (10.10) to allow enough roots to fit both the decay and the oscillation for each component. Thus, in fitting a two-component curve, one does not use the Prony's method for two components but rather determines the values of U from the Prony method as if one were looking for four components. The result is usually that one finds two positive values of U and two negative values. One can then determine the values of the A's by using only the positive roots of U and calculating two values of A.

In the example shown, the four values of U were 0.90, 0.50, -0.60, -0.20. The two positive values were used to find the values of A of 86 and 8.9.

In the general case, to fit a curve that is believed to consist of K exponential components, one applies the Prony method as if one were looking for $2K$ components, and finds $2K$ values of U of which K will be positive. One then calculates K values of A, using the positive values of U.

The other difficulty associated with the Prony method, and in fact with any exponential stripping method, is determining the correct number of components. It is always possible to improve the fit by increasing the number of components into which the curve is to be separated. As a result, one has difficulty defining the proper number of components. If the solution to a problem is to be reasonable, it must be relatively insensitive to small variations in the data. We cannot accept, as a reasonable solution to a problem, one that changes radically for very small changes in the data. A reasonable test for this is to take a set of data and divide it into two subsets consisting of alternate points. Thus, if the data points are numbered 1 through N, we take as one subset of data, points 1, 3, 5, 7, ..., and as the other subset points 2, 4, 6, 8, ... Each of these is then separately analyzed by the Prony method and the results compared. If these results agree with each other fairly well, we can assume that we have not attempted to find too many components, but may not have enough. If they disagree badly, we must reduce the number of components into which we are trying to separate the data curve.

As an example, the two subsets of data in Table X-1, which are alternate points from a very smooth two-component curve, are analyzed as both two components and three components, using the modified Prony method.

As two components, the two sets yielded

Data set A $\quad Y = 79.9(0.498)^n + 10.0(0.899)^n$

Data set B $\quad Y = 80.0(0.500)^n + 10.0(0.900)^n$

But as three components, one finds

Data set A $\quad Y = 78.5(0.496)^n + 9.96(0.899)^n + 1.48(0.60)^n$

Data set B $\quad Y = 80.2(0.500)^n + 10.0(0.899)^n - 0.157(0.018)^n$

Note the wide discrepancy between the A and B sets in the third component. This indicates that the apparent third component is a result of statistical fluctuation. Since no such discrepancy was found in the two-component separation, we would conclude (correctly) that the curve was made up of two components.

Table X-1

n	A	B
0	90.0	—
1	—	49.0
2	28.1	—
3	—	17.3
4	11.5	—
5	—	8.40
6	6.56	—
7	—	5.41
8	4.62	—
9	—	4.03
10	3.56	—
11	—	3.18
12	2.48	—
13	—	2.55
14	2.29	—
15	—	2.06
16	1.85	—
17	—	1.67
18	1.50	—
19	—	1.35
20	1.21	—
21	—	1.09
22	0.984	—
23	—	0.886

9. Inverted Exponentials

The Prony method can also be applied to inverted exponentials or a combination of inverted and decaying exponentials by the following scheme. Let

$$\begin{aligned} Y_0 &= A_0 + A_1 + A_2 \\ Y_1 &= A_0 + A_1 U_1 + A_2 U_2 \\ Y_2 &= A_0 + A_1 U_1{}^2 + A_2 U_2{}^2 \\ Y_N &= A_0 + A_1 U_1{}^N + A_2 U_2{}^N \end{aligned} \tag{10.11}$$

As before, construct a new set of equations by multiplying groups of three of the foregoing by 1, M_1, M_2, and add each group.

$$\begin{aligned} Y_0 + M_1 Y_1 + M_2 Y_2 = A_0(1 + M_1 + M_2) &+ A_1(1 + M_1 U_1 + M_2 U_1{}^2) \\ &+ A_2(1 + M_1 U_2 + M_2 U_2{}^2) \end{aligned}$$

$$Y_1 + M_1 Y_2 + M_2 Y_3 = A_0(1 + M_1 + M_2) + A_1 U_1(1 + M_1 U_1 + M_2 U_1^2) + A_2 U_2(1 + M_1 U_2 + M_2 U_2^2)$$

$$\vdots$$

$$Y_{N-2} + M_1 Y_{N-1} + M_2 Y_N = A_0(1 + M_1 + M_2) + A_1 U_1^{N-2} \times (1 + M_1 U_1 + M_2 U_1^2) + A_2 U_2^{N-2}(1 + M_1 U_2 + M_2 U_2^2)$$

Once again, the values of U are chosen to be the roots of

$$1 + M_1 U + M_2 U^2 = 0 \tag{10.12}$$

so that the brackets containing U are zero and

$$Y_0 + M_1 Y_1 + M_2 Y_2 = A_0(1 + M_1 + M_2)$$

$$Y_1 + M_1 Y_2 + M_2 Y_3 = A_0(1 + M_1 + M_2)$$

$$Y_{N-2} + M_1 Y_{N-1} + M_2 Y_N = A_0(1 + M_1 + M_2)$$

The foregoing are not simultaneous linear equations in the M's because A_0 is a function of the M's. However, by defining a new variable M_0, we can make them resemble simultaneous linear equations. Let

$$M_0 = A_0(1 + M_1 + M_2)$$

Then

$$\begin{aligned} Y_0 + M_1 Y_1 + M_2 Y_2 - M_0 &= 0 \\ Y_1 + M_1 Y_2 + M_2 Y_3 - M_0 &= 0 \\ &\vdots \\ Y_{N-2} + M_1 Y_{N-1} + M_2 Y_N - M_0 &= 0 \end{aligned} \tag{10.13}$$

In general, we will have more such equations than unknowns, so that they cannot all be equal to zero for fixed values of the M's. We therefore use the method of least squares to minimize

$$S = \sum_{J=0}^{N-2} (Y_J + M_1 Y_{J+1} + M_2 Y_{J+2} - M_0)^2$$

by finding the partial derivatives with respect to each M:

$$\frac{\partial S}{\partial M_1} = 2 \sum_{J=0}^{N-2} Y_{J+1}(Y_J + M_1 Y_{J+1} + M_2 Y_{J+2} - M_0)$$

$$\frac{\partial S}{\partial M_2} = 2 \sum_{J=0}^{N-2} Y_{J+2}(Y_J + M_1 Y_{J+1} + M_2 Y_{J+2} - M_0)$$

$$\frac{\partial S}{\partial M_0} = -2 \sum_{J=0}^{N-2} (Y_J + M_1 Y_{J+1} + M_2 Y_{J+2} - M_0)$$

Setting these equal to zero, we find the normal equations

$$-M_0 \sum Y_{J+1} + M_1 \sum Y_{J+1}^2 + M_2 \sum Y_{J+1} Y_{J+2} + \sum Y_J Y_{J+1} = 0$$
$$-M_0 \sum Y_{J+2} + M_1 \sum Y_{J+1} Y_{J+2} + M_2 \sum Y_{J+2}^2 + \sum Y_J Y_{J+2} = 0$$
$$-M_0[N-1] + M_1 \sum Y_{J+1} + M_2 \sum Y_{J+2} + \sum Y_J = 0 \tag{10.14}$$

which are solved for M_1, M_2, and M_0. The values of U are then determined from (10.12) and inserted into (10.11) to find the values of A by the method of least squares. Once again, the same consideration of straddle points applies, and one must usually double the number of components to allow for negative values of U.

The ultimate test of the validity of an exponential separation is whether or not it can consistently distinguish between an experiment and its control. One can argue endlessly over the validity of exponential separation, and in fact, mathematically, it is somewhat tenuous, but if the result consistently distinguishes the experiment from the control, then whether or not the components represent physical entities is not relevant. We accomplish what we set out to accomplish: the analysis of an experiment. Except in those cases where the compartments are physically or chemically distinguishable (of which there are, incidentally, many), we must be careful about assigning physical meaning to the components. One way of looking at this is to assume that one is not trying to assign physical meaning but rather trying to find a compact way of describing the shape of a decay curve.

10. The General Method for Nonlinear Least-Squares Fitting

The method to be described is useful for fitting a wide variety of nonlinear functional relationships, including multiple exponentials. Although more general than the Prony method, we shall see that it suffers from at least one major drawback, which is that it is necessary to have some idea of the optimum values of the parameters to use as starting values.

Stated in its most general terms, the problem of nonlinear least-squares fitting in one independent variable is the following. Given a set of values of the independent variable X_J, where J runs from 1 to N, and the corresponding values of the dependent variable Y_J or $Y(X_J)$, we wish to minimize the sum of the squares of the residuals

$$S = \sum_{J=1}^{N} W_J[f(A_K, X_J) - Y(X_J)]^2 \tag{10.15}$$

where $f(A_K, X_J)$ is a given function of M parameters $A_1, A_2, \ldots, A_M$ and the independent variables X_J. The W are weighting factors, which may be different for each point or may all have the value one. The choice of the functional relationship is usually based on some characteristic of the problem as discussed in Section 5, but is occasionally determined by trying a variety of relationships and finding the one that produces the smallest value of the sum of residuals squared. In the remainder of this chapter, we shall assume that the weighting factors are all unity, and that the functional form $f(A_K, X_J)$ is known.

In the earlier case of linear least-squares fitting, we generated a set of normal equations by differentiating the sum of the squared residuals with respect to each parameter. These normal equations were then a set of simultaneous linear equations, which are easily solved. In the nonlinear case, the derivatives may be difficult to calculate, the resulting normal equations are not likely to be linear, and the solution may be very difficult. One is tempted to try the following strategy. Assume a set of reasonable starting values of the parameters. Calculate the sum of the squared residuals resulting from using these values. Now try changing one of the parameter values slightly, and recalculate the sum of the squared residuals. If it has become smaller, continue to change this parameter in the same direction. When it no longer gets smaller, try changing a different parameter. Alternatively, one might modify this strategy by trying a change in A_1 that is found to improve the sum squared residuals, followed by a change in A_2 that is found to improve it; followed by a change in A_1 again. Unfortunately, this rarely works. What frequently happens is that one gets trapped in boxes. A new value of A_1 is found which has improved the sum squared residuals. A further improvement is found by changing A_2. When one goes back to change A_1 again to see if this will result in still more improvement, one finds that to improve the sum squared residuals by changing A_1 requires restoring it in the direction of its initial value. When this is tried, it is found that the best change in A_2 is one that moves it toward its initial value, and the process repeats cyclically, never yielding significantly improved values of the sum squared residuals. The situation is analogous to a chess game in which White's queen has checked Black's king. Black moves his king. White moves his queen to a new position, which checks Black's king again. Black moves his king back to its original position, White moves his queen to its original position. Black is in check again. Black moves, White moves, and so on. The process goes on endlessly. The game is a draw. Clearly, what White must do is to try some new move that has not yet been explored. In the least-squares case, the new move rarely improves the situation. The process gets trapped in the same sort of ring. The only thing for White to do is to move all of his pieces simultaneously, which, though illegal in chess, is quite legal in mathematics, and we shall now show how one chooses to change all of the parameters simultaneously to avoid the foregoing trap.

(In the following we have replaced A_1 and A_2 by A and B to simplify the notation.) Suppose we have the relatively simple relationship

$$Y(X) = AX^2 + BX$$
$$Y(1) = 5, \qquad Y(2) = 14, \qquad Y(3) = 26 \tag{10.16}$$

and let us assume that we have some reason to believe that the appropriate values for A and B are, respectively, somewhere near 1 and 2. Let us calculate residuals from these values. We find

$$\begin{aligned} R_1 &= Y(1) - (A \cdot 1^2 + B \cdot 1) = 2 \\ R_2 &= Y(2) - (A \cdot 2^2 + B \cdot 2) = 6 \\ R_3 &= Y(3) - (A \cdot 3^2 + B \cdot 3) = 11 \end{aligned} \tag{10.17}$$

Now, ideally we ought to be able to adjust the values of A and B in such a way as to make these residuals smaller. In order to do so, we have to know how each residual will change if we change A and B by the amounts ΔA and ΔB. Residual 1 will change as follows

$$\Delta R_1 = \frac{\partial R_1}{\partial A} \Delta A + \frac{\partial R_1}{\partial B} \Delta B$$

while residuals 2 and 3 will change by

$$\Delta R_2 = \frac{\partial R_2}{\partial A} \Delta A + \frac{\partial R_2}{\partial B} \Delta B, \qquad \Delta R_3 = \frac{\partial R_3}{\partial A} \Delta A + \frac{\partial R_3}{\partial B} \Delta B$$

Now, how much do we want these residuals to change? We would like the change in R_1 to reduce its current value, 2, and we would like the change in R_2 to exactly remove its value, 6, and the change in R_3 to reduce its value by 11. Therefore, we set

$$\Delta R_1 = -2 = \frac{\partial R_1}{\partial A} \Delta A + \frac{\partial R_1}{\partial B} \Delta B$$

$$\Delta R_2 = -6 = \frac{\partial R_2}{\partial A} \Delta A + \frac{\partial R_2}{\partial B} \Delta B$$

$$\Delta R_3 = -11 = \frac{\partial R_3}{\partial A} \Delta A + \frac{\partial R_3}{\partial B} \Delta B$$

We now have a set of simultaneous linear equations in ΔA and ΔB. In this case, it happens that we can very easily compute the partial derivatives from (10.17):

$$\frac{\partial R_1}{\partial A} = -1, \qquad \frac{\partial R_1}{\partial B} = -1$$

$$\frac{\partial R_2}{\partial A} = -4, \qquad \frac{\partial R_2}{\partial B} = -2$$

$$\frac{\partial R_3}{\partial A} = -9, \qquad \frac{\partial R_3}{\partial B} = -3$$

Thus,

$$\Delta A + \Delta B = 2$$
$$4\,\Delta A + 2\,\Delta B = 6$$
$$9\,\Delta A + 3\,\Delta B = 11$$

These three equations cannot all be satisfied simultaneously, since the solution of the first two is $\Delta A = 1, \Delta B = 1$, which is not a solution to the third equation. On the other hand, it is close to a solution; so, rather than apply the method of least squares at this point, which we usually do when we have this situation, let us just go ahead and try $\Delta A = 1$, $\Delta B = 1$. The new values of A and B, which we call A' and B', are

$$A' = A + \Delta A = 2, \qquad B' = B + \Delta B = 3$$

and the new residuals are

$$R_1' = Y(1) - (2 \cdot 1^2 + 3 \cdot 1) = 0$$
$$R_2' = Y(2) - (2 \cdot 2^2 + 3 \cdot 2) = 0$$
$$R_3' = Y(3) - (2 \cdot 3^2 + 3 \cdot 3) = -1$$

which are clearly much better than the first set (10.17). We could now repeat the process, using the method of least squares this time, to get a still further improved set of values for A and B.

The foregoing example was particularly simple in that we were able to compute the partial derivatives easily. This was the result of having chosen a very simple function (10.16). Had we chosen a more complicated function, one whose derivatives were not easily calculated, we would have had to find partial derivatives numerically. This is quite easy to do. To find the partial derivative of $f(A, B, X_J)$ with respect to A, given starting values of A and B, we introduce a small change in A, compute a new value of f, and say that the partial derivative must be approximately

$$\frac{\partial f(A, B, X_J)}{\partial A} \approx \frac{f(A + \Delta A, B, X_J) - f(A, B, X_J)}{\Delta A}$$

We then restore A to its initial value, and try changing B slightly to find a partial derivative with respect to B. This is all done as a purely numerical process, and as long as our function f is a reasonably smooth function of A and B, the partial derivatives thus calculated will be very close to exact values.

Let us develop the theory of the general case of nonlinear curve fitting. We assume that we have a function $f(A_K, X_J)$. Corresponding to each value of J, from 1 to N, we have an X_J and a Y_J, which is a data point. The A_K take the place of A and B in the example above and we use the notation A_K to indicate that we may have more than two parameters to be fit. For example, if we have M parameters, K would go from 1 to M. For each data point we construct a residual,

$$R_J = Y_J - f(A_K, X_J)$$

and now we compute how this residual changes if we change each of the parameters, each A_K, in turn. The net change in each residual is given by

$$\Delta R_J = \sum_{K=1}^{M} \Delta A_K \frac{\partial f(A_K, X_J)}{\partial A_K}, \qquad J = 1, \ldots, N$$

(note that this expression represents N different equations, one for each value of J) and we would like these changes to be equal to the negatives of the residuals we have computed from our starting values.

$$\Delta R_J = -R_J$$

In general, we will not be able to do this exactly, so let us construct a new set of residuals, which we shall call errors, as we did earlier in this chapter.

$$E_J = R_J + \Delta R_J$$

Our goal now is to minimize the sum of the squares of these errors

$$S' = \sum_{J=1}^{N} (E_J)^2 = \sum_{J=1}^{N} \left(R_J + \sum_{K=1}^{M} \Delta A_K \frac{\partial f(A_K, X_J)}{\partial A_K} \right)^2$$

which we do by differentiating S' with respect to each of the ΔA

$$\frac{\partial S'}{\partial(\Delta A_L)} = 2 \sum_{J=1}^{N} \left(R_J + \sum_{K=1}^{M} \Delta A_K \frac{\partial f}{\partial A_K} \right) \frac{\partial f}{\partial A_L}$$

and setting each derivative equal to zero. This yields M normal equations

$$\sum_{J=1}^{N} \sum_{K=1}^{M} \frac{\partial f(A_K, X_J)}{\partial A_L} \frac{\partial f(A_K, X_J)}{\partial A_K} \Delta A_K + \sum_{J=1}^{N} R_J \frac{\partial f(A_K, X_J)}{\partial A_L} = 0,$$

$$L = 1, 2, 3, \ldots, M$$

which appear hopelessly complicated, but remember that the partial derivatives that occur here are numerical values, which we have found by calculating for each X_J the values of f for changes in each A in turn; thus, though the notation is complicated, the result is fairly simple. We have constructed a set of normal equations just as we did in the linear least-squares method, which we solve for the ΔA_K. We now add the ΔA_K to the starting values and repeat the process, using the new value of A_K, which we call A_K'. If we are lucky, the sum of squared errors at this stage will be less than it was initially, and we repeat the process, making it consistently smaller and smaller, until we get no noticeable change. The only real problem is that of finding an appropriate set of starting values for the A_K. As in the one-dimensional case, which we did in Section 1, we shall find that inappropriate starting values will lead us into traps from which we cannot recover. Even worse, unlike the one-dimensional example, we shall not always know when we are in these traps because we shall find a set of sum squared errors that is smaller than those we found from our starting values but we shall have no way, in general, of knowing whether it is the *least possible* sum of squared errors. There is, fortunately, a test for this, which is to start from varieties of different values of the A_K. If we find that the process consistently converges to the same final values of the A_K, we may be reasonably sure that we have been going down the right hill. On the other hand, if we find that the different sets of starting values yield different results, we are being trapped in the wrong valleys.

In many problems, the problem itself will suggest appropriate starting values of the parameters and, with a little bit of luck the method described earlier will then improve these to the optimum parameters. In many other cases, however, one needs some simple strategy by which to find appropriate starting parameters. There are a great variety of these; in particular, we wish to point out that a very good set of starting values for exponential curves can be obtained from the Prony method, outlined earlier in this chapter.

The foregoing process is not too complicated in principle. In practice, the amount of numerical computation required is horrendous! Consequently, digital computers are invariably used to solve problems of this type, and prepared programs exist for the purpose. The programs used generally have built into them a number of features that are not required by the foregoing theory, but are quite useful in the solution of practical problems. For example, one can usually insert limit values of each of the parameters, values that, for some physical reason, cannot be exceeded by the correct solution. In this way the computer can interrupt the solution of a problem if it is getting off on a hopeless track. Another trick that is often incorporated is to check the values of the sum squared residuals after each correction of the parameters to be sure that it is really getting smaller. The theory above indicates that it should get smaller, but occasionally the process gets into difficulty from the approximation of the differential by a straight line approximation to the derivative. In cases where this does not hold, the computer program tries a smaller step. Still more elaborate strategy can be incorporated, but we leave these to the ingenuity of the computer programmers.

Tracer Experiments

1. Introduction: Quantification of Radioactivity

Tracers are one of the principal tools of biological and particularly of biochemical research. Although the range of situations in which they are used is vast, the three types of usage illustrated in the following examples display many of the principles, and some of the problems, that arise.

There are many kinds of tracers in use, but by far the most common are radioactively labeled compounds. We need, therefore, to have some familiarity with the terminology used to quantify radioactivity. The measures most frequently used in the biological laboratory are the microcurie and the millicurie, which are 3.7×10^4 and 3.7×10^7 disintegrations per second, respectively. Although it is possible to measure the actual number of disintegrations per second in a sample, more often one counts some fixed but unknown fraction of the actual disintegration rate. The reason that only a fraction is counted has to do with the experimental artifacts in the counting devices, that is, absorption of photons by containers, internal quenching, or imperfect efficiency of the counter. In order to avoid a dependence upon absolute counting rates, we shall assume that all measurements are made under the same conditions, which need not be specified precisely, and that the observed counting rate is equal to an efficiency factor E multiplied by the absolute sample radioactivity. This will allow us to interpret all measurements in terms of relative count rates, and as long as the counting configuration, and therefore the efficiency, remains the same, we need not refer to absolute count rates.

In biological literature, one encounters the quantity called *specific activity*, which can be defined in various ways but is usually given in terms of radioactivity per unit mass of sample. Specific activity provides a simple and easily remembered way of stating certain physical laws, but its use in developing the mathematics of the movement of tracers can obscure the physical principles involved. Throughout this chapter, therefore, we shall indicate quantities of radioactive compound with the same symbol as is used for the nonradioactive compound, but with the addition of an asterisk, and we shall work with ratios of these quantities. Conversion to specific activity will be done only when some definite advantage can be obtained from that form.

2. Flux through Cell Membranes

One of the interesting properties of cells is the rate at which various compounds or ions enter or leave them. To measure the influx, which is the rate at which a substance enters a cell, cells that are free of radioactivity are placed in a bath containing the compound or ion to be measured, some fraction of which is radioactively labeled. The time at which the cells are placed into the radioactive medium will be referred to as time zero. At various times thereafter, a quantity of the mixture of medium and cells is removed and the cells separated from the medium by centrifugation. The amount of radioactivity that has been taken up by the cells is measured and plotted as a function of the length of time they were in the radioactive medium.

The measurement of outflow from cells, efflux, is done by reversing this process. Cells are grown in the medium containing the radioactive substance to be measured. At time zero, the cells are transferred to a nonradioactive medium, and at some later time they are separated from this by centrifugation. In this case, one measures the appearance of radioactivity in the previously unlabeled medium.

In principle, both of these measurements are very simple. In practice, there are difficulties in the interpretation of both. One problem is that during centrifugation, a certain amount of radioactive substance is trapped on the outside of the cells and in the space between cells, even though the cells are tightly packed against the bottom of the centrifuge tube. Thus the radioactivity of the sedimented cell pellet includes both radioactivity internal to the cells and that which is stuck to the outside of the cells. In the case of an efflux measurement, when the cells are separated from the radioactive growing medium and resuspended in the nonradioactive medium, this extracellular radioactivity appears in the medium almost instantaneously as an initial base value, so that the washout curve as a function of time does not extrapolate to zero at time zero. Even washing the cells between the time they are labeled and the time they are measured does not fully solve this problem. The wash time must be severely limited or the labeled substance to be measured will diffuse out of the cell, thereby confusing measurements of the true uptake; and limited washes do not succeed in removing all of the extracellular material. In influx measurements a similar difficulty occurs. In the following analysis we shall see how one corrects for this type of experimental artifact.

Chapter V discussed the mathematics of diffusion and active transport of compounds between biological compartments. Throughout that chapter, in order to separate mathematical complexities from biological complexities, it was assumed that the physical parameters L and K characterizing the partition separating the compartments were known and of fixed value. In many experimental situations one is attempting to determine these quantities. One wishes to find the rate at which substances are transported into or out of cells from measurements of these physical quantities, which in Chapter V appeared in the solution of problems. For example, using the equation following (5.2), one could measure $C(t)$ or $VC(t)$ and thereby determine I and L.

Solutions of the type described in Chapter V cannot simply be reversed. Such naive reversal to determine the unknown parameters without due consideration to artifacts such as extracellular radioactivity will lead to grossly incorrect results. The current task is to investigate realistic circumstances and show how to correct for experimental artifacts in a number of simple cases.

One major simplifying assumption is made in this section, which is that the measuring process being studied does not itself change the values of the parameters being measured. Thus, for example, if one is studying the movement of an electrically charged ion across a cell membrane, one assumes that this does not significantly change the electrical potential difference between the inside and the outside of the cell, which would result in a time-dependent variation of the physical parameters; the complexities thus introduced are far beyond the scope of this text. Fortunately, in a great number of cases this approximation is valid, but one must always keep in mind that it has been made.

3. Efflux Measurements

For efflux measurements, cells are grown in a medium containing a radioactively labeled variety of the compound to be measured and are transferred at time zero to an otherwise identical but nonradioactive medium. At some later time, the cells are separated from the nonradioactive medium by centrifugation and the radioactivity remaining in the cells, or that appearing in the previously nonradioactive medium, is measured. Normally radioactivity within the cell would be expected to fall exponentially as a function of time.

In practice, this is difficult to observe because of two common complications. The first is that for many compounds, the amount of radioactive compound leaving the cell during the experiment is a very small fraction of that initially incorporated into the cell, and therefore one can only observe a short part of the washout curve. The second is that the radioactive growing medium can adhere to the outside of the cell, so that one does not directly measure the washout curve, but rather the washout curve plus a fixed amount of radioactivity outside of the cells.

In order to observe the washout curve, one runs a number of parallel samples. The reason for this is that the process of separating the cells from the nonradioactive medium for the purpose of measuring their radioactivity terminates the experiment. Therefore each sample can be used to determine the percentage washout only at one value of time. In order to get the entire curve, one must start a number of samples simultaneously and terminate each at a different time. Of course the samples must initially be as much alike as possible. The usual procedure is to grow the cells in a single radioactive bath and separate them into samples of equal size which are then measured separately.

A compound diffusing out of a cell at a flux rate Φ_{OUT} carries with it labeled compound in the ratio of labeled compound to the total of labeled plus unlabeled, which is approximately just the total compound. Thus the appropriate differential equation for the washout of the labeled compound is

$$\frac{dQ^*}{dt} = -\Phi_{\mathrm{OUT}} \frac{Q^*}{Q} \tag{11.1}$$

where * indicates the radioactive component and Q is the quantity of the compound in the cell. Thus

$$Q^*(t) = Q^*(0)e^{-(\Phi_{\mathrm{OUT}}/Q)t} \tag{11.2}$$

The appearance of radioactivity in the supernate is given by

$$Q^*_{\mathrm{SUP}}(t) = nQ^*(0)(1 - e^{-(\Phi_{\mathrm{OUT}}/Q)t}) \tag{11.3}$$

where n is the number of cells in the sample.

It is often difficult to obtain enough data to define the curve described by (11.3). The time during which some kinds of cells are viable in culture is often short relative to the time constant of the exponential term. When this is true, one takes advantage of the fact that in a sufficiently dilute supernate, eventually all of the radioactivity within the cell pellet will appear in the supernate; and since the cell pellet can be counted by itself, the counting efficiency factor E applies to both and therefore does not appear in the final result. Thus one may determine the flux rate of the compound in question from the following, in which Ω is the observed count rate:

$$\Omega_{\text{PELLET}} = EQ^*_{\text{PELLET}} = EnQ^*(0)$$
$$\Omega_{\text{SUP}} = EQ^*_{\text{SUP}}(t) = EnQ^*(0)(1 - e^{-(\Phi_{\text{OUT}}/Q)t})$$
$$\frac{\Omega_{\text{SUP}}}{\Omega_{\text{PELLET}}} = 1 - e^{-(\Phi_{\text{OUT}}/Q)t} \tag{11.4}$$
$$\frac{d}{dt}\left(\frac{\Omega_{\text{SUP}}}{\Omega_{\text{PELLET}}}\right)\bigg|_0 = \frac{\Phi_{\text{OUT}}}{Q}$$

In principle, one needs only to determine the time derivative of the supernate count rate with respect to the pellet rate at time zero to get the ratio of flux rate to compound within the cell pellet. Unfortunately the trapped extracellular material makes this a difficult quantity to determine. Furthermore, for small time intervals, one cannot usually get a sufficient difference in count rates to determine the slope of the curve with any accuracy. One must therefore modify the above procedure to incorporate the effects of the trapped extracellular component and to allow the use of sufficiently different time values to yield a reliably measurable change in the count rate of the supernate. Let the count rate be measured at a sequence of times t_1 through t_n, indicated as t_j below, measured from the time at which the cells are separated from the radioactive medium. Thus we have the following:

$$\Omega_{\text{SUP}}(t_j) = \Omega_{\text{TRAPPED}} + EnQ^*(0)(1 - e^{-(\Phi_{\text{OUT}}/Q)t_j}) \tag{11.5}$$

where Ω_{TRAPPED} represents that part of the observed count rate due to extracellular material. But one also knows what the count rate at infinite time would be, even though one may not be able to measure it directly due to limitations of cell life. At infinite time, all of the radioactive material would have diffused out of the cell, and the supernate count rate would therefore equal the pellet count rate:

$$\Omega_{\text{SUP}}(\infty) = \Omega_{\text{PELLET}} = \Omega_{\text{TRAPPED}} + EnQ^*(0) \tag{11.6}$$

One can now remove the unknown quantities E, n, and $Q^*(0)$ from this equation and plot the ratio shown below on semilogarithmic paper, for which one should get a straight line intercepting the time origin close to, but not at, the value unity:

$$\begin{aligned}\frac{\Omega_{\text{PELLET}} - \Omega_{\text{SUP}}(t_j)}{\Omega_{\text{PELLET}}} &= \frac{EnQ^*(0)}{\Omega_{\text{PELLET}}}\, e^{-(\Phi_{\text{OUT}}/Q)t_j} \\ &= \frac{(\Omega_{\text{PELLET}} - \Omega_{\text{TRAPPED}})}{\Omega_{\text{PELLET}}}\, e^{-(\Phi_{\text{OUT}}/Q)t_j} \\ &= \left(1 - \frac{\Omega_{\text{TRAPPED}}}{\Omega_{\text{PELLET}}}\right) e^{-(\Phi_{\text{OUT}}/Q)t_j}.\end{aligned} \tag{11.7}$$

The difference between the value unity and the observed value at time zero is the ratio of the counts of the trapped component to the total pellet count, and the quantity one needs to determine, Φ_{OUT}/Q, can be found from the slope of this line.

4. Influx Measurements

For influx, one measures the rate at which radioactivity is accumulated within cells that have been grown in a nonradioactive medium and then transferred to a radioactive medium. The analysis of this type of measurement is complicated by the fact that as soon as radioactive isotope appears within the cell, it begins to diffuse out as well. By the time one has enough isotope in the cell for an accurate measurement, it is necessary to account for the quantity leaving the cell through outward-directed flux. Thus the equation for the rate of accumulation of radioactive material in a cell is

$$\frac{dQ^*}{dt} = \Phi_{IN}\frac{C^*_{EXT}}{C_{EXT}} - \Phi_{OUT}\frac{C^*_{INT}}{C_{INT}} = V\frac{dC^*_{INT}}{dt} \qquad (11.8)$$

where C_{EXT} represents concentrations outside of the cell, C_{INT} concentrations inside of the cell, V is the volume of the cell, and * indicates the radioactive component.

The inward-directed component of flux can be found by measuring the rate of change of internal radioactivity, but one must measure close to time zero because, as we shall see, it is not possible to derive the necessary information from the shape of the uptake curve. One might expect that the shape of the curve would allow one to determine the influx, but counterintuitive though it may seem, the shape of the curve is actually determined by the efflux constant. One may make use of this to get an independent determination of the efflux constant, but it does not help one to find the influx. The only direct measurement one can get of influx is by measuring the rate of change of radioactive content near time zero. If one happens to know that the cells are in a steady state with respect to the compound being measured, one can use the efflux constant to measure influx, but if one does not know this, as one might not in some experiments, there is no choice but to obtain a sequence of measurements at small time values which will allow one to estimate the slope at time zero.

As in the efflux experiment, when attempting the measurement by centrifuging a pellet from a hot medium, one may have to contend with extracellular trapped radioactivity. The best one can do is to take a number of points closely spaced near time zero, extrapolate to time zero, and assume that this extrapolation yields the trapped component. The influx constant is then found in terms of the observed count rate in the pellet and the count rate of an aliquot of radioactive medium in the following manner:

$$\left.\frac{d\Omega_{\text{PELLET}}}{dt}\right|_0 = En\left.\frac{dQ^*}{dt}\right|_0 = \Phi_{\text{IN}}\frac{C^*_{\text{EXT}}}{C_{\text{EXT}}}En$$

$$\Omega_{\text{ALIQUOT}} = EC^*_{\text{EXT}}V_{\text{ALIQUOT}} \tag{11.9}$$

$$\Phi_{\text{IN}} = \frac{C_{\text{EXT}}V_{\text{ALIQUOT}}}{n\Omega_{\text{ALIQUOT}}}\left.\frac{d\Omega_{\text{PELLET}}}{dt}\right|_0$$

On occasion there is difficulty in measuring $d\Omega_{\text{PELLET}}/dt|_0$. If one measures the slope between time points t_1 and t_2, both different from zero, one can use as an approximation

$$\left.\frac{d\Omega}{dt}\right|_{\frac{t_1+t_2}{2}} = \left.\frac{d\Omega}{dt}\right|_0 e^{-(\Phi_{\text{OUT}}/Q)(t_1+t_2)/2} \tag{11.10}$$

In some cases one knows Φ_{OUT}/Q from efflux experiments. In other cases one can use an approximate $d\Omega/dt|_0$ to get an approximate Φ_{IN}, and assuming a steady state, use this value as Φ_{OUT} in the correction shown above. This may appear to be reasoning in a circle, but the exponential factor is not much different from unity, so that the correction though not perfect is adequate. Of cardinal importance, however, is the accurate definition of the time values involved. Careless laboratory techniques which introduce errors in the measurement of the starting time will cause gross errors in the correction factor.

5. Survival Time Measurements

Professor Pike Poisson was an avid fisherman who lived by the mysterious deep pond of Florida. One day Professor Poisson decided to determine how many fish there were in the pond and what their average lifetime was. To do this he caught 1,000 fish, labeled them by crimping little metal tags to their tails, and returned them to the pond.

A few days later, after sufficient time to ensure that his tagged fish were thoroughly mixed with the other fish in the pond, Professor Poisson caught another 1,000 fish and of these he observed that 1% had tags on their tails. He then returned the 1,000 fish to the pond. Since 1% had tags on their tails and he assumed thorough mixing, he was able to infer that 1% of all the fish in the pond had tags on their tails, and since he knew there were 1,000 tagged fish, he could infer that there were 100,000 fish in the pond.

A year later Professor Poisson repeated the second part of his experiment. He again caught 1,000 fish but this time found that only 9/10 of 1% of them had tags on their tails. Assuming a stable population in the pond, Professor Poisson concluded that 1/10 of his labeled fish had died during the year, and since the tagged fish were a representative population of the fish in the pond when first caught, he concluded that 1/10 of all the fish in the pond die each year. From this he arrived at the fact that the average lifetime of a fish in the pond is 10 years.

Just to check his result, Professor Poisson repeated his assay in each of the next two years, but at the end of the second year he found that slightly more than 10% of his tagged fish had vanished during that year. At the end of the third year he found that a still slightly larger fraction of the tagged fish had vanished in the third year than in the second year. This puzzled him for a few minutes, but then he realized that the number of tagged fish left could not decay in a simple ratio from year to year, since such a decay rate would imply a decaying exponential population function which never reaches zero. He knew that at some time after he had originally tagged his 1,000 fish all of them would have died, and therefore the curve of the remaining tagged fish as a function of time could not be a simple decaying exponential.

The correct explanation of the phenomenon Professor Poisson observed is as follows. The initial catch of 1,000 fish was presumably a representative population of the fish in the pond at that time. At the end of two years, however, the labeled fish were on the average older than a representative sample would have been and so were no longer a representative population. The correct method of analyzing the experiment, which we are sure Professor Poisson quickly understood, is to extrapolate the curve of remaining tagged fish versus time back to time zero, preferably on semilogarithmic paper where an almost straight line will result for the first few years. In later years the line will curve downward. The slope of the semilogarithmic plot of the early years yields the death rate of a representative sample.

6. Precursor Product Relationships

A common problem of biochemistry is to establish the order in a sequence of biochemical reactions. Consider the following: Substance A is converted into substance B which in turn is converted to substance C. One may ask the following questions. Does A go directly to B or is there an intermediate compound A' formed? Does all of A get converted to B or is some of A converted to something else? Another possibility is that B is formed from A but also from F. Yet another possibility is that not all of B becomes C. And so on.

In this section we are going to make use of kinetic principles to resolve these questions. In each case we shall assume that the compounds A and B can be obtained in pure form and labeled with radioactive isotope, so that they can be introduced into the system and their appearance in subsequent pools can be determined as a function of time. A problem one encounters in this process is that of recovering pure compounds from the pool. For example, if one is observing the formation of B from A, it is not usually possible to recover all of B. In fact, the more specifically the test for B separates it from other compounds, the greater the difficulty encountered in recovering a known fraction of the B pool. However, we can avoid having to recover all of B or even a known fraction of B by formulating results in terms of specific activity or in terms of relative count rates for samples and standards.

We assume as an initial model the following: A is converted to B which in turn is converted to C. A might also be converted simultaneously to D and B to G (Figure 11-1).

We shall also make the following assumptions:

1. The pool sizes are stable.
2. The addition of a tracer sample of A does not significantly change the pool.
3. All reactions are in one direction.
4. The radioactive sample of A has been prepared in a manner that guarantees that it is pure A and that the radioactive sample introduced mixes with the A pool instantly.
5. The radioactive part of A is handled biologically in a manner indistinguishable from that of the existing nonradioactive component.

The radioactively labeled sample of A is introduced into the A pool at time zero. At subsequent times, samples of A from the pool are extracted. The count rate observable from a fixed quantity of labeled A and that determined from the same quantity of A recovered from the pool are then determined.

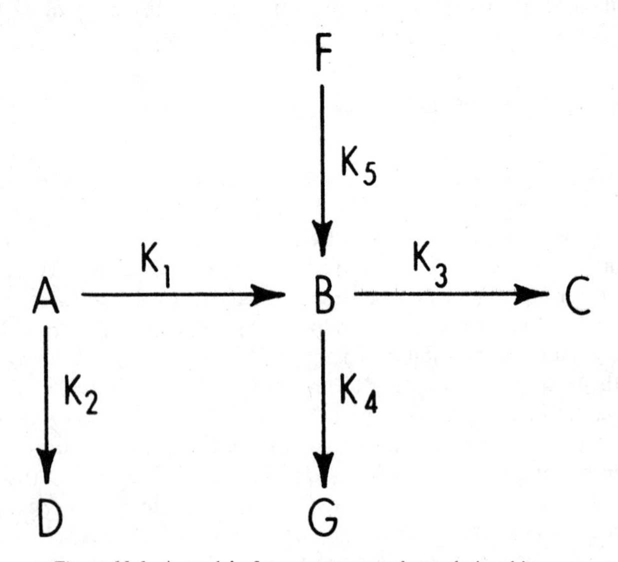

Figure 11-1. A model of a precursor product relationship.

This ratio allows us to determine the size of the A pool and the sum of the two rate constants K_1 and K_2. The ratio r of count rates is plotted semilogarithmically as a function of time. The value of r found by extrapolating the ratio as a function of time to time zero allows us to determine the size of the A pool:

$$r(0) = \frac{A^*}{A} \tag{11.11}$$

$$A = \frac{A^*}{r(0)} \tag{11.12}$$

Since A^* will vanish from the A pool exponentially, the count rate determined from samples of the A pool plotted semilogarithmically against time will determine the sum of the rate constants K_1 and K_2, a quantity we shall call α:

$$\frac{dA^*}{dt} = -(K_1 + K_2)A\frac{A^*}{A} = -(K_1 + K_2)A^*$$

$$A^*(t) = A^*(0)e^{-(K_1+K_2)t} \tag{11.13}$$

$$\frac{r(t)}{r(0)} = e^{-(K_1+K_2)t} = e^{-\alpha t}$$

We now label a sample of B, which we shall call B^{**} to emphasize that this is not radioactivity entering B from A. We can then determine the size of the B pool and the sum of the two outward-directed rate constants K_3 and K_4, the sum of which we shall call β. The steady-state equations based on the assumption of stable pool sizes and the temporary assumption that the input from F is zero are

$$AK_1 = B(K_3 + K_4) = \beta B$$

$$K_1 = \beta\frac{B}{A} \tag{11.14}$$

From the two experiments described above, we can find whether α is equal to K_1 or is more than K_1, and thus whether all of A is converted into B. Similarly, if we find that β, the sum of K_3 and K_4, is inconsistent with the rate of known input from A, we can infer that, contrary to the assumption above, there is some source of B other than A, such as F in Figure 11-1.

The above experiment, though conceptually simple, may be fairly hard or even impossible to perform because of the difficulty of getting a radioactive sample of B sufficiently well-mixed with the B pool. In this case we must resort to a different tactic in which we follow radioactivity from A into B. We set up the appropriate rate equation for B^* (single asterisk to indicate that it is coming from A):

$$\frac{dB^*}{dt} = K_1A^* - \beta B^* = K_1A^*(0)e^{-\alpha t} - \beta B^* \tag{11.15}$$

and solve in the standard way to yield:

$$B^*(t) = A^*(0)\frac{K_1}{\alpha - \beta}(e^{-\beta t} - e^{-\alpha t}) \tag{11.16}$$

We now evaluate the following quantities, which are suggested by analogy to measurements of blood flow and vascular volume:

$$\begin{aligned} \int_0^\infty B^*(t)dt &= \frac{A^*(0)K_1}{\alpha\beta} \\ \int_0^\infty tB^*(t)dt &= A^*(0)K_1\frac{\alpha + \beta}{\alpha^2\beta^2} \end{aligned} \tag{11.17}$$

With a little bit of manipulation we find:

$$\frac{\int_0^\infty tB^*(t)dt}{\int_0^\infty B^*(t)dt} = \frac{\alpha + \beta}{\alpha\beta} = \frac{1}{\alpha} + \frac{1}{\beta} \tag{11.18}$$

Since we know α, we can determine β. Once again we compare α and β determined from isotopes with the relationships known for the steady state, and interpret discrepancies as the existence of other compartments.

Another question one might ask is whether A is the immediate precursor of B. One obvious clue is that B^* should start to appear immediately following the injection of A^*. This, however, may be difficult to determine because of the time delays inherent in administering and withdrawing samples. Another clue which is a little easier to interpret is the shape of the appearance curve of B^*. If A is the immediate precursor of B and A^* is known to be vanishing in an ever-decaying exponential, the rate of appearance of B^* must decrease. Therefore the leading edge of the B^* versus time curve should be downward curved. It is also useful to look at the point in time at which the activity of the B pool peaks. If A is the immediate precursor of B, we can easily show that the specific activity of A is equal to the specific activity of B at the point in time at which B^* peaks. First, we note that

$$\frac{dB^*}{dt} = K_1A^*(t) - \beta B^*(t) \tag{11.19}$$

Now B^* peaks when $dB^*/dt = 0$ or when

$$K_1A^* = \beta B^*$$

But since the pool sizes are stable,

$$K_1A = \beta B$$

Therefore,

$$\frac{A^*}{A} = \frac{B^*}{B} \text{ at the time } B^* \text{ reaches its peak.} \tag{11.20}$$

If this relationship is not found to be true, one must assume that the simple model on which it is based is not appropriate.

APPENDIX I

The Rules of Differentiation

To prove that the derivative of the product of a constant and a function of x is the constant multiplied by the derivative of the function, let

$$y(x) = Cf(x)$$

Then

$$y(x + \Delta x) = Cf(x + \Delta x)$$

and

$$\frac{dy}{dx} = \underset{\Delta x \to 0}{\text{limit}} \frac{y(x + \Delta x) - y(x)}{\Delta x}$$

$$= C \underset{\Delta x \to 0}{\text{limit}} \frac{f(x + \Delta x) - f(x)}{\Delta x} = C \frac{df}{dx}$$

To prove that the derivative of a sum of two functions is the sum of their individual derivatives, let

$$y(x) = f(x) + g(x)$$

Then

$$y(x + \Delta x) = f(x + \Delta x) + g(x + \Delta x)$$

$$\frac{dy}{dx} = \underset{\Delta x \to 0}{\text{limit}} \frac{y(x + \Delta x) - y(x)}{\Delta x}$$

$$= \underset{\Delta x \to 0}{\text{limit}} \frac{f(x + \Delta x) - f(x)}{\Delta x} + \frac{g(x + \Delta x) - g(x)}{\Delta x}$$

$$= \frac{df}{dx} + \frac{dg}{dx}$$

To show that the derivative of a product of two functions is the first multiplied by the derivative of the second plus the second multiplied by the derivative of the first, let

$$y(x) = f(x)g(x)$$

$$y(x + \Delta x) = f(x + \Delta x)g(x + \Delta x)$$

$$= \left(f(x) + \Delta x \frac{df}{dx}\right)\left(g(x) + \Delta x \frac{dg}{dx}\right)$$

$$\frac{y(x + \Delta x) - y(x)}{\Delta x} = g(x)\frac{df}{dx} + f(x)\frac{dg}{dx} + \Delta x \frac{dg}{dx}\frac{df}{dx}$$

$$\operatorname*{limit}_{\Delta x \to 0} \frac{y(x + \Delta x) - y(x)}{\Delta x} = g(x)\frac{df}{dx} + f(x)\frac{dg}{dx} = \frac{dy}{dx}$$

To show that the derivative of the quotient of two functions is the denominator multiplied by the derivative of the numerator minus the numerator multiplied by the denominator, all divided by the square of the denominator, let

$$y(x) = \frac{f(x)}{g(x)}$$

$$y(x + \Delta x) = \frac{f(x + \Delta x)}{g(x + \Delta x)} = \frac{f(x) + \Delta x \dfrac{df}{dx}}{g(x) + \Delta x \dfrac{dg}{dx}}$$

$$\frac{y(x + \Delta x) - y(x)}{\Delta x} = \frac{1}{\Delta x}\left[\frac{f(x)g(x) + \Delta x g(x)\dfrac{df}{dx} - f(x)g(x) - \Delta x f(x)\dfrac{dg}{dx}}{g^2(x) + \Delta x g(x)\dfrac{dg}{dx}}\right]$$

$$\frac{dy}{dx} = \operatorname*{limit}_{\Delta x \to 0} = \frac{g(x)\dfrac{df}{dx} - f(x)\dfrac{dg}{dx}}{g^2(x)}$$

1. The Derivative of e^x

Hopefully, the reader remembers from elementary algebra the existence of the binomial theorem, which states that one can express the nth power of a sum of a and b as

$$(a + b)^n = a^n + \frac{n}{1} a^{n-1}b + \frac{n(n-1)}{1 \cdot 2} a^{n-2}b^2$$

$$+ \frac{n(n-1)(n-2)}{1 \cdot 2 \cdot 3} a^{n-3}b^3 + \cdots + \frac{n(n-1)(n-2)\cdots 1}{1 \cdot 2 \cdot 3 \cdot 4 \cdots n} b^n$$

The rules for this series are

1. It consists of $n + 1$ terms beginning with $a^n b^0$ ($b^0 = 1$) and each succeeding term is formed from the previous term by reducing the exponent of a by one and increasing the exponent of b by one until the last term, which contains $a^0 b^n = b^n$.

2. The first term is preceded by the coefficient 1. The coefficient of each succeeding term is found from the previous coefficient by multiplying it by the exponent of its a term and dividing it by the exponent of b increased by one. Thus if the kth term is

$$C_k a^p b^q$$

the next term is

$$C_{k+1} a^{p-1} b^{q+1}$$

where

$$C_{k+1} = C_k \frac{p}{q+1}$$

Consider the quantity

$$E^x = [(1 + \delta)^{1/\delta}]^x = (1 + \delta)^{x/\delta}$$

where δ is a number considerably less than one. For any given value of x it is possible to find a value for E^x for any value of δ provided that x/δ is an integer n. According to the binomial theorem,

$$E^x = (1 + \delta)^n = 1 + \frac{n}{1}\delta + \frac{n(n-1)}{1 \cdot 2}\delta^2$$

$$+ \frac{n(n-1)(n-2)}{1 \cdot 2 \cdot 3}\delta^3 + \cdots + \frac{n(n-1)(n-2)\cdots 1}{1 \cdot 2 \cdot 3 \cdot 4 \cdots n}\delta^n$$

Consider the Jth term of this series; it begins with n and has $J - 1$ factors containing n in the numerator

$$\frac{n(n-1)(n-2)\cdots(n-J+2)}{1 \cdot 2 \cdot 3 \cdots (J-1)}\delta^{J-1}$$

Thus the $J = 5$ term is

$$\frac{n(n-1)(n-2)(n-3)}{1 \cdot 2 \cdot 3 \cdot 4} \delta^4$$

Now

$$\delta^4 = \frac{x^4}{n^4}$$

and the $J = 5$ term becomes

$$\frac{(1)\left(1 - \frac{1}{n}\right)\left(1 - \frac{2}{n}\right)\left(1 - \frac{3}{n}\right)}{1 \cdot 2 \cdot 3 \cdot 4} x^4$$

which for n sufficiently large becomes

$$\frac{x^4}{4!}$$

where the notation 4! means the product $1 \cdot 2 \cdot 3 \cdot 4$. Thus

$$E^x \approx 1 + \frac{x}{1!} + \frac{x^2}{2!} + \frac{x^3}{3!} + \cdots + \frac{x^n}{n!}$$

From the way in which E^x was defined, it is known to be some number E raised to the x power. The number E is found by setting x equal to one.

$$E = 1 + 1 + \frac{1}{2!} + \frac{1}{3!} + \cdots + \frac{1}{n!}$$

So far there is nothing unusual about this series. For given values of x and n the sum is easily computed. We shall now show that the sum can be computed for infinite n. More precisely, we shall show that it is possible to find a number E_{lower}, which will be less than the exact value of E, and another number E_{upper}, which will be greater than the exact E, and that these two numbers E_{lower} and E_{upper} can be brought very close together, so that the exact value of E is trapped between them.

Let E_n be the sum of the series for a specific n. Let E_{n_1} be the sum for another n_1 greater than n.

$$E_{n_1} = \left(1 + 1 + \frac{1}{2!} + \frac{1}{3!} + \cdots + \frac{1}{n!}\right) + \left(\frac{1}{(n+1)!} + \cdots + \frac{1}{n_1!}\right)$$

Clearly $E_{n_1} > E_n$.

Thus E_n for any n becomes a lower limit for E as n becomes infinite. We can now compute E_n for various n and the largest such value of n for which we choose to do the computation becomes E_{lower}.

$$E_1 = 1$$
$$E_2 = 1 + 1 = 2$$
$$E_3 = 1 + 1 + \frac{1}{2} = 2.5$$
$$E_4 = 1 + 1 + \frac{1}{2} + \frac{1}{6} = 2.66$$
$$E_{20} = 1 + \cdots + \frac{1}{20!} = 2.7182$$

The harder problem is to show that there is some number E_{upper} that is larger than E and to find E_{upper} sufficiently close to some E_{lower} to fix E between E_{lower} and E_{upper}. To do so we return to E_n^x

$$E_n^x = 1 + x + \frac{x^2}{2!} + \frac{x^3}{3!} + \cdots + \frac{x^n}{n!}$$

$$E_n^x - 1 = x + \frac{x^2}{2!} + \frac{x^3}{3!} + \cdots + \frac{x^n}{n!}$$

$$x(E_n^x - 1) = x^2 + \frac{x^3}{2!} + \frac{x^4}{3!} \cdots \frac{x^n}{(n-1)!} + \frac{x^{n+1}}{n!}$$

also

$$E_n^x - 1 - x = \frac{x^2}{2!} + \frac{x^3}{3!} + \frac{x^4}{4!} + \cdots + \frac{x^n}{n!} + 0$$

Note that each term in the bottom line is less than the corresponding term in the line just above it. Therefore,

$$E_n^x - 1 - x < x(E_n^x - 1) \quad \text{or} \quad E_n^x - 1 < xE_n^x$$

Add the quantity $1 - E_n^x$ to both sides of this inequality.

$$E_n^x - xE_n^x < 1$$

Divide both sides by $1 - x$

$$E_n^x < \frac{1}{1-x}$$

Note that this is true for any n and x, so that by calculating one such value for a given x we have an upper bound for E, which we call E_{upper}.

If we let $x = \frac{1}{2}$, then

$$E_n^{1/2} < \frac{1}{1 - \frac{1}{2}} = 2, \qquad E_n < 4$$

This is not a very good upper bound. It is too far from E_{lower}. If we try $x = 0.1$, then

$$E_n^{0.1} < \frac{1}{1 - 0.1} = 1.1111$$

We now raise each side to the tenth power:

$$E_n < (1.1111)^{10} = 2.865$$

If we continue both these processes, calculating better values for E_{lower} by making n bigger and better values for E_{upper} by making x smaller, we find $E = 2.7182$, which is the number called e.

We now have the result

$$e^x = 1 + \frac{x}{1!} + \frac{x^2}{2!} + \frac{x^3}{3!} + \cdots + \frac{x^n}{n!}$$

if n is very very large. We shall now show that

$$\frac{d}{dx} e^x = e^x$$

Let us differentiate the foregoing series term by term.

$$\frac{d}{dx} e^x = 0 + \frac{1}{1!} + \frac{2x}{2!} + \frac{3x^2}{3!} + \cdots + \frac{nx^{n-1}}{n!}$$

$$= 1 + \frac{x}{1!} + \frac{x^2}{2!} + \cdots + \frac{x^{n-1}}{(n-1)!}$$

This series is exactly the same as the series for e^x except that e^x contains one more term $x^n/n!$. In order to show that this last term $x^n/n!$ is negligible compared to e^x, let us compare it to one of the middle terms of e^x.

The term in the middle of the series (for n even) is

$$\frac{x^{n/2}}{(n/2)!}$$

Now let us compute the ratio between the last term and this middle term.

$$R = \frac{x^{n/2}/(n/2)!}{x^n/n!} = \frac{x^{n/2}(n/2)!}{n!}$$

$$\frac{(n/2)!}{n!} = \frac{1}{\left(\frac{n}{2} + 1\right)\left(\frac{n}{2} + 2\right) \cdots (n-1)n}$$

where there are $n/2$ factors in the denominator. Now

$$\left(\frac{n}{2}+1\right)\left(\frac{n}{2}+2\right)\cdots(n-1)n > \left(\frac{n}{2}\right)^{n/2}$$

Therefore,

$$R < \frac{x^{n/2}}{(n/2)^{n/2}} = \left(\frac{2x}{n}\right)^{n/2}$$

Now choose n greater than $2x$ and raised to an infinite power as $n \to \infty$, R becomes zero. But e^x is certainly greater than its middle term. Therefore, e^x is much greater than the last term, which can therefore be neglected. The proof is complete.

From the foregoing we may also derive the derivative of a logarithm. The reader will recall that the logarithm of a number x is defined as the power to which a base number must be raised to equal x. Thus, if B is the base and y the log of x,

$$x = B^y, \qquad y = \log_B x$$

In this book all logarithms are taken to the base e, so that

$$x = e^y, \qquad y(x) = \ln_e x = \ln x$$

To find the derivative, calculate

$$y(x + \Delta x) = \ln(x + \Delta x)$$
$$x + \Delta x = e^{y+\Delta y} = e^y e^{\Delta y} = xe^{\Delta y}$$

If Δy is small,

$$e^{\Delta y} = 1 + \Delta y$$

Therefore,

$$x + \Delta x = x(1 + \Delta y) = x + x\Delta y$$

$$\frac{\Delta y}{\Delta x} = \frac{1}{x}$$

$$\frac{dy}{dx} = \operatorname*{limit}_{\Delta x \to 0} \frac{\Delta y}{\Delta x} = \frac{1}{x}, \qquad \int \frac{1}{x}\,dx = \ln x$$

APPENDIX II

Table of Laplace Transforms

	Inverse transform ←	Laplace transform →	
	$G(t)$	$g(s)$	
1.	1	$\frac{1}{s}$	
2.	t	$\frac{1}{s^2}$	
3.	t^2	$\frac{2}{s^3}$	
4.	t^n	$\frac{n!}{s^{n+1}}$	
5.	e^{-at}	$\frac{1}{s+a}$,	$a \neq 0$
6.	$\frac{1}{a}(1-e^{-at})$	$\frac{1}{s(s+a)}$,	$a \neq 0$
7.	$\frac{1}{a}\left[t-\frac{1}{a}(1-e^{-at})\right]$	$\frac{1}{s^2(s+a)}$,	$a \neq 0$
8.	te^{-at}	$\frac{1}{(s+a)^2}$,	$a \neq 0$

Inverse transform ←		Laplace transform →	
	$G(t)$	$g(s)$	
9.	$(1-at)e^{-at}$	$\dfrac{s}{(s+a)^2}$,	$a \neq 0$
10.	$\dfrac{1}{a^2}[1-(at+1)e^{-at}]$	$\dfrac{1}{s(s+a)^2}$,	$a \neq 0$
11.	$\dfrac{1}{a-b}(e^{-bt}-e^{-at})$	$\dfrac{1}{(s+a)(s+b)}$,	$a \neq b,\quad a \neq 0,\quad b \neq 0$
12.	See Chapter IV, Section 10	$\dfrac{1}{s^2+ps+q}$	
13.	See Chapter IV, Section 10	$\dfrac{s}{s^2+ps+q}$	
14.	See Chapter IV, Section 10	$\dfrac{1}{s(s^2+ps+q)}$	
15.	$\dfrac{1}{2a}(e^{at}-e^{-at})$	$\dfrac{1}{s^2-a^2}$,	$a \neq 0$
16.	$\frac{1}{2}(e^{at}+e^{-at})$	$\dfrac{s}{s^2-a^2}$,	$a \neq 0$
17.	$\dfrac{1}{2a^2}(e^{at}+e^{-at}-2)$	$\dfrac{1}{s(s^2-a^2)}$,	$a \neq 0$
18.	$\dfrac{1}{ab}\left(1+\dfrac{be^{-at}-ae^{-bt}}{a-b}\right)$	$\dfrac{1}{s(s+a)(s+b)}$	
19.	$\dfrac{ae^{-at}-be^{-bt}}{a-b}$	$\dfrac{s}{(s+a)(s+b)}$	
20.	$\dfrac{1}{a}\sin at$	$\dfrac{1}{s^2+a^2}$	
21.	$\cos at$	$\dfrac{s}{s^2+a^2}$	

APPENDIX III

Derivatives of Sine and Cosine Functions

The derivatives of the sine and cosine functions are most easily found by means of a geometrical construction. For the cosine function, consider the two triangles shown in Figure AIII-1. If R is the radius of a circle, the two horizontal legs are, respectively, $R\cos\theta$ and $R\cos(\theta + \Delta\theta)$. The length of the arc connecting the tops of the triangles is approximately a straight line, and is of length $R\,\Delta\theta$ if the angle is defined in radians (this point being true because the full circle has length $2\pi R$ and there are 2π radians in a circle). The length difference between the two horizontal legs is, in terms of the straight line approximation to the arc, equal to $R\,\Delta\theta\sin\theta$. Therefore,

$$R[\cos(\theta + \Delta\theta) - \cos\theta] = -R\,\Delta\theta\sin\theta$$

$$\frac{d}{d\theta}\cos\theta = \underset{\Delta\theta\to 0}{\text{limit}}\ \frac{\cos(\theta + \Delta\theta) - \cos\theta}{\Delta\theta} = -\sin\theta$$

The proof for the sine function is analogous.

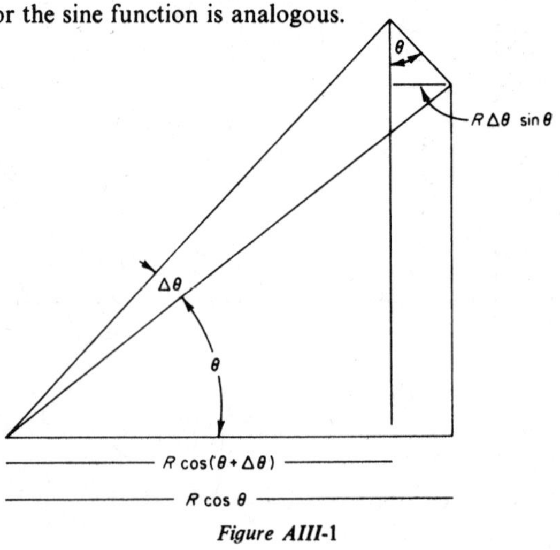

Figure AIII-1

APPENDIX IV

Solutions to Selected Problems

Chapter I

1. $\lim_{\Delta x \to 0} \dfrac{G(x + \Delta x) - G(x)}{\Delta x}$

2. $W(2) = C, \quad \dfrac{dW(2)}{dx} = 2$

3. By the product rule,

$$\frac{dU(2)}{dx} = W(2)\frac{dV(2)}{dx} + V(2)\frac{dW(2)}{dx}$$

$$\frac{dU(2)}{dx} = 3 \times 2 + 4 \times 4 = 22$$

By the quotient rule,

$$\frac{dS(2)}{dx} = \frac{W(2)\frac{dV(2)}{dx} - V(2)\frac{dW(2)}{dx}}{W^2(2)} = \frac{3 \times 2 - 4 \times 4}{3^2} = -\frac{10}{9}$$

4. $\dfrac{dy}{dx}$ is approximately $\dfrac{Y(3.1) - Y(3.0)}{0.1} = \dfrac{0.3}{0.1} = 3$

5. $y(x + \Delta x) = y(x) + \Delta x\dfrac{dy}{dx}$

$$y(3.1) \approx y(3) + 0.1 \times 2 = 12.2$$

6. $\dfrac{dV}{dx} = \dfrac{dg}{du}\dfrac{df}{dx}$

7. $\dfrac{dV}{dx} = \dfrac{dg}{du}\dfrac{du}{dx} = 8x^7 + 16x^3$

8. $C(t) = 2e^{+kt}$ moles/liter

9. $C(t) = 2e^{k(t-1)}$ moles/liter

10. $Ae^{-k4} = \frac{1}{2}Ae^{-k2}$, $e^{2k} = 2$

From Figure 1-6, $2 = e^{0.7}$ approximately, $k = 0.35$ sec^{-1}.

11. Decay half time is 1 sec. Therefore,

$e^{-\alpha 1} = \frac{1}{2}$

From (1.15),

$e^{-0.69} = \frac{1}{2}$, $\alpha = 0.69$ sec^{-1}

12. $e^{-2.3} = \dfrac{1}{10}$, $e^{-\alpha\tau_{1/10}} = \dfrac{1}{10}$, $\alpha = \dfrac{2.3}{\tau_{1/10}}$

13. (a) K in sec^{-2}

(b) K in liter^2/mole

(c) K in liters/sec

14. 180.39×10^{10} furlongs/fortnight

15. $$\frac{\text{moles/liter}}{\text{sec}} = \frac{\text{moles/sec}}{\text{liter}} - \frac{\text{liters/sec}}{\text{liter}}\frac{\text{moles}}{\text{liter}}$$

16. $$\frac{\text{moles/liter}}{\text{sec}^2} + B\frac{\text{moles/liter}}{\text{sec}} + D\frac{\text{moles}}{\text{liter}} = E$$

B must therefore have units of sec^{-1}; D, sec^{-2}; E, moles liter^{-1} sec^{-2}.

17. (a) $4x^3 + 9x^2 + 1$ (b) $4x/(x^2 + 1)^2$

(c) $2xe^{x^2}$ (d) $(1 + \alpha x)e^{\alpha x}$

Chapter II

1.

Time (sec)	Q (moles)
0	0.300
1	0.294
2	0.288
3	0.282
4	0.276

3. In a small interval of time Δt the change of Q in the cell is

$\Delta Q = + \Delta t K(C_E - C_C)$

$$\Delta C_C = \frac{+\Delta Q}{V} = +\Delta t \frac{K}{V}(C_E - C_C)$$

Divide by Δt and take the limit as Δt goes to zero.

$$\frac{dC_C}{dt} = +\frac{K}{V}(C_E - C_C), \quad C_C(0) = \frac{Q(0)}{V}$$

4. $\Delta Q = \Delta t\, K(C_E - C_C) + R\Delta t, \qquad \frac{dC_C}{dt} = \frac{K}{V}(C_E - C_C) + \frac{R}{V}$

5. $C(0) = \frac{Q(0)}{V}$ where $C(0) = 0.024$ moles/liter, $Q(0) = 0.1$ moles

   Solve to find $V = 4.167$ liters.

   To find F, plot data points on semilogarithmic paper. The points do not lie in a straight line due to experimental error, but one may draw an approximate line, determine its half range, and solve for F, which is found to be approximately 2.61 liters/sec.

6. Since the half range is 24 hours, $-0.69 = 24k$.

   Solve for k and evaluate e^{-kt} for $t = 10$ hours.

   Three-fourths of the original quantity remains after 10 hours.

7. $\frac{dC_3(t)}{dt} = \frac{F}{V_3}(C_2(t) - C_3(t))$ where $C_2(t)$ is given by (2.14).

8. $A(0) = 0, \qquad \frac{dA(t)}{dt} = R - KA(t)$

9. $\frac{dA}{dt} = R - KA + K'B, \qquad \frac{dB}{dt} = KA - K'B$

10. $y_1 = 2e^{-x}, \qquad y_3 = 2(1 - e^{-x})$

Chapter III

2. $B^2 = 4D$

3. $A = 4, \qquad B = 0$ (See Eq. (3.5).)

4. $\frac{A - 5}{A + B} \stackrel{?}{=} \frac{e^{-5t}}{t^2}$

   This relation cannot be true for all values of t unless

   $A - 5 = 0, \qquad A + B = 0, \qquad A = 5, \qquad B = -5$

5. (a) $C(0) = 5, \qquad \frac{dC}{dt} = K_1 t + 4$

(b) $C(0) = 3, \quad \frac{dC}{dt} = K_1 t + 5$

6. $\frac{dC(t)}{dt} = K_1 t + A_1$

$$\frac{d^2C(t)}{dt^2} = K_1 \neq K_1 C(t) \text{ unless } K_1 = 0$$

7. $\frac{dC_2}{dt} = \frac{dC_1}{dt}, \quad \frac{d^2C_2}{dt^2} = \frac{d^2C_1}{dt^2}$

Therefore,

$$\frac{d^2C_2}{dt^2} + K_1\frac{dC_2}{dt} + K_2C_2 = \frac{d^2C_1}{dt^2} + K_1\frac{dC_1}{dt} + K_2\left(C_1 + \frac{K_3}{K_2}\right)$$

and since C_1 is a solution of

$$\frac{d^2C_1}{dt^2} + K_1\frac{dC_1}{dt} + K_2C_1 = 0$$

the previous equation is equal to K_3.

8. Use the relationship: If

$$C_3 = C_1 + C_2$$

then

$$\frac{d^2C_3}{dt^2} + K_1\frac{dC_3}{dt} + K_2C_3 = \frac{d^2C_1}{dt^2} + K_1\frac{dC_1}{dt} + K_2C_1$$

$$+ \frac{d^2C_2}{dt^2} + K_1\frac{dC_2}{dt} + K_2C_2 = 0$$

since C_1 and C_2 are both solutions of

$$\frac{d^2C}{dt^2} + K_1\frac{dC}{dt} + K_2C = 0$$

9. (a) Linear, homogenous, first order; (b) nonlinear, first order; (c) linear, inhomogenous, second order; (d) nonlinear, third order.

10. (a), (c), (d).

11. (a) $C = 2e^{4t}$.

(c) From entry 4 of Table III-1 we have

$$C = A_1t + A_2, \quad C(1) = A_1 + A_2 = 2, \quad C(2) = 2A_1 + A_2 = 3$$

$$A_1 = 1, \quad A_2 = 1$$

12. $\dfrac{dC(t)}{dt} = \alpha_1 A_1 e^{\alpha_1 t} + \alpha_2 A_2 e^{\alpha_2 t}$

$$\frac{d^2C(t)}{dt^2} = \alpha_1^2 A_1 e^{\alpha_1 t} + \alpha_2^2 A_2 e^{\alpha_2 t}$$

13. α_1 is not equal to α_2.

14. Yes. $C_2(t) = \dfrac{F}{V_2} C_1(0)\, te^{-(F/V_2)t}$

17. 20 minutes

18. 1.5 μCi

19. No. Insulin production is not linearly related to blood sugar content.

Chapter IV

1. $\dfrac{26}{6}$ cubic centimeters

2. $\dfrac{V}{F}$ sec

3. $A = 4/3, \quad B = -4/3, \quad C = 0.$ Yes, C makes sense.

4. $x = 1, \quad y = 2, \quad z = 3$

5. $x = \dfrac{CE - BF}{AE - BD}$

If C and F are zero, x and y are both zero unless

$$AE = BD$$

in which case unique solutions for x and y cannot be found. In this case any pair x, y for which

$$\frac{x}{y} = -\frac{B}{A} = -\frac{E}{D}$$

will satisfy both of the given equations.

6. $y = 3e^{kt}$

7. $y(t) = \dfrac{-1}{k^2T^2}(1 - e^{kt}) + 2e^{kt} - \dfrac{t}{kT^2}$

8. $C(t) = \dfrac{A}{\alpha - (L/V)}(e^{-(L/V)t} - e^{-\alpha t}) + \dfrac{R}{L}(1 - e^{-(L/V)t})$

9. $C(t) = \frac{1}{2}(e^{at} + e^{-at})$

10. $C(t) = \frac{1}{2a}(e^{at} - e^{-at})$

11. $Y_1 = \frac{R}{K}\sin Kt, \qquad Y_2 = \frac{-R}{K}\cos Kt + \frac{R}{K}$

12. $x(p) = \frac{0.9\, e^{-0.1p} - 0.1\, e^{-0.9p}}{0.8}$

Chapter V

1. As usual, the first thing one should do with a problem of this type is to try to sketch the anticipated result. It is clear from the statement of the problem that after the pump stops operating the internal concentration must become equal to the external concentration. While the pump is operating, internal concentration must be less than external. One should immediately suspect that an inverted exponential is the solution. While the pump is operating, the rate at which the pumped and diffusing substance enters the cell is

$$\frac{dQ}{dt} = V\frac{dC_{\text{in}}}{dt} = -R - K(C_{\text{in}} - C_{\text{ext}})$$

which in the steady state yields for the internal concentration

$$C_{\text{ss}} = C_{\text{ext}} - \frac{R}{K}$$

Therefore, the inverted exponential must start at C_{ss} and go to C_{ext}. Let us try an exponential form that has this property.

$$C_{\text{in}}(t) = C_{\text{ss}}e^{-\alpha t} + C_{\text{ext}}(1 - e^{-\alpha t})$$

After the pump stops,

$$\frac{dC_{\text{in}}}{dt} = -\frac{K}{V}(C_{\text{in}} - C_{\text{ext}})$$

Let us differentiate the proposed solution and see if it can be made to fit:

$$\frac{dC_{\text{in}}}{dt} = -\alpha C_{\text{ss}}e^{-\alpha t} + \alpha C_{\text{ext}}e^{-\alpha t} = -\alpha e^{-\alpha t}(C_{\text{ss}} - C_{\text{ext}})$$

$$\stackrel{?}{=} -\frac{K}{V}(C_{\text{in}} - C_{\text{ext}}) = -\frac{K}{V}e^{-\alpha t}(C_{\text{ss}} - C_{\text{ext}})$$

If $K/V = \alpha$, the solution fits.
Alternatively, one could have used the Laplace transform.

$$sg(s) - C_{in}(0) = -\frac{K}{V}\left(g(s) - \frac{C_{ext}}{s}\right)$$

$$g(s) = \frac{C_{in}(0)}{s + \frac{K}{V}} + \frac{K}{V}\frac{C_{ext}}{s\left(s + \frac{K}{V}\right)}$$

$$C_{in}(0) = C_{ss}$$

$$C(t) = C_{ss}e^{-(K/V)t} + C_{ext}(1 - e^{-(K/V)t})$$

2. While the pump is operating,

$$(C_{ext} - C_{in})K = R, \qquad C_{in} = C_{ext} - \frac{R}{K}$$

After the pump stops, this problem is identical to that described in Chapter III, Section 10.

Let V_1 be the volume inside the cell, V_2 the outside volume.

$$C_1(0) = C_{in} = C_{ext}(0) - \frac{R}{K}$$

$$C_2(0) = C_{ext}(0)$$

$$C_1(t) = C_{ext}(0) - \frac{R}{K} + \frac{R}{K}(1 - e^{-\alpha t})\frac{V_2}{V_1 + V_2}$$

$$\alpha = K\frac{V_1 + V_2}{V_1 V_2}$$

3. We make use of the convolution equation (5.31).

$$C(t) = \int_0^t I(\tau)H(t - \tau)d\tau$$

For a single compartment

$$H(t) = \frac{1}{V}e^{-(F/V)t}, \qquad H(t - \tau) = \frac{1}{V}e^{-(F/V)(t-\tau)}$$

$I(t)$ is given as

$$I(t) = I_0e^{-(F/V)t}, \qquad I(\tau) = I_0e^{-(F/V)\tau},$$

$$C(t) = \frac{1}{V}\int_0^t I_0e^{-(F/V)\tau}e^{-(F/V)(t-\tau)}d\tau$$

$$= \frac{1}{V}\int_0^t I_0e^{-(F/V)t}\,d\tau = \frac{1}{V}I_0te^{-(F/V)t}$$

4. Case 1:

$$\frac{dC_2(t)}{dt} = \frac{-K}{V_2}[C_2(t) - C_1(t)]$$

$$\frac{dC_1(t)}{dt} = \frac{K}{V_1}[C_2(t) - C_1(t)] - \frac{F}{V_1}C_1(t)$$

$$\frac{dC_2(0)}{dt} = 0$$

Then solve by Laplace transform.

Case 2:

$$C(t) = C_0 e^{-[F/(V_1+V_2)]t}$$

5. The output function consists of the line segment $I_0 t$ for $0 \leq t \leq 1$, the line segment I_0 for $1 \leq t \leq 10$, and the line segment $(11-t)I_0$ for $10 \leq t \leq 11$. The function is zero for all other values of t.

6. $$\frac{dC_1}{dt} \approx \frac{1}{V_1}[I_1 - (L_1 + K)C_1]$$

$$C_1(0) = 0$$

Solving this, we find

$$C_1(t) = \frac{I_1}{L_1 + K}(1 - e^{-t(L_1+K)/V_1})$$

Thus

$$\frac{dC_2}{dt} + \frac{L_2}{V_2}C_2 = \frac{K}{V_2}\,\frac{I_1}{L_1 + K}(1 - e^{-t(L_1+K)/V_1})$$

Solve this by the Laplace transform. You should find that the steady state is given by

$$C_2(\infty) = \frac{KI_1}{L_2(L_1 + K)}$$

7. $C(0) = 0, \quad C(1) = 0, \quad C(2) = 16, \quad C(3) = 28, \quad C(4) = 10, \quad C(n) = 0$ for all $n \geq 5$.

8. $$C_{2ss} = \frac{I_1 K}{L_1 L_2 + K(L_1 + L_2)}, \qquad C_{1ss} = \frac{I_1(L_2 + K)}{L_1 L_2 + K(L_1 + L_2)}$$

Chapter VII

1. $V_{ZF} = 1003\frac{1}{3}\text{ cm}^3, \qquad V \geq 1000\text{ cm}^3$

2. $\Delta V = -\frac{1}{3}, \qquad \Delta V = -1$

3. $\frac{dV}{dt} = G(V_{ZF} - V) - KV$

   $V_{ZF} = 1033\frac{1}{3}\text{ cm}^3$

   If Q is added at time zero, $V(t) = 1000 + Qe^{-3.1t}$

4. $V_{ZF} = 996\frac{2}{3}\text{ cm}^3$

5. Let $V' = V_2(t) - V_2(0)$.

   $$\frac{d^2V'}{dt^2} + (K_1 + K_2)\frac{dV'}{dt} + K_1(G + K_2)V' = 0, \qquad V'(0) = 0,$$

   $$\frac{dV'(0)}{dt} = K_1Q$$

Chapter VIII

2. The derivation of (8.25) starts from Eq. (8.21) and its accessory conditions with L equal to zero:

   $$\frac{d^2C}{dx^2} = \frac{R}{K}, \qquad C(0) = 0, \qquad \frac{dC(0)}{dx} = 0$$

   Apply the Laplace transform, with which we find

   $$s^2g(s) - sC(0) - \frac{dC(0)}{dx} = \frac{R}{sK}, \qquad g(s) = \frac{R}{s^3K}$$

   This is inverted by means of entry 4, Appendix II, with n equal to 2:

   $$C(x) = \frac{Rx^n}{Kn!} = \frac{Rx^2}{K2}$$

3. The formulation of this problem is

   $$\frac{d^2C}{dx^2} = \frac{L}{K}C, \qquad C(0) \text{ given}$$

with the additional condition that C must remain finite for all x. We solve this by means of the Laplace transform:

$$s^2 g(s) - sC(0) - \frac{dC(0)}{dx} = \frac{L}{K} g$$

$$\left(s^2 - \frac{L}{K}\right) g = sC(0) + \frac{dC(0)}{dx}$$

$$g = \frac{sC(0)}{s^2 - (L/K)} + \frac{dC(0)}{dx} \frac{1}{s^2 - (L/K)}$$

$$C(x) = \frac{1}{2}\left(C(0) + \frac{1}{\alpha}\frac{dC(0)}{dx}\right) e^{\alpha x} + \frac{1}{2}\left(C(0) - \frac{1}{\alpha}\frac{dC(0)}{dx}\right) e^{-\alpha x}$$

$$\alpha = \sqrt{\frac{L}{K}}$$

If C is to remain finite, the coefficient of the first term must be zero; therefore

$$\frac{1}{\alpha}\frac{dC(0)}{dx} = -C(0) \qquad \text{and} \qquad C(x) = C(0)e^{-\alpha x}$$

4. $\frac{d^2C}{dx^2} = \frac{L}{K} C_T e^{x\sqrt{L/K}} = \frac{L}{K} C(x)$

5. If $L = 0$, $C(x) = \frac{R}{2K}(x^2 - D^2) + C(D)$.

   If $I = 0$, use (8.34).

6. The value of the series is zero at the end points.

8. (a) The requirement is that the inside temperature reaches half that of the surface. This will occur when

$$\sum_{n=1}^{\infty} (-1)^n e^{-(n\pi/r_0)^2 Kt} = -\tfrac{1}{4}$$

Now

$$e^{-0.69} = 0.50$$

therefore

$$e^{-1.38} = 0.25$$

Looking at the first term of the series by itself, we see

$$\left(\frac{\pi}{r_0}\right)^2 Kt = 1.38$$

and since

$$K = 2\frac{69}{\pi^2}\frac{\text{cm}^2}{\text{sec}} \quad \text{and} \quad r_0 = 10 \text{ cm}$$

then

$t = 1$ sec

The remaining terms of the series can be neglected. For example, at t equal to 1 sec the second term of the series is

$$e^{-1.38\times 4} = e^{-5.52} = 0.004$$

which is negligible compared to 0.25.

(b) The product Kt must have the same value as in part (a). Therefore,

$t = \frac{1}{2}$ sec.

(c) t/r_0^2 must have the same value as in part (a), and since r_0 is doubled,

$t = 4$ sec.

9. It is assumed that $C(r)$ is positive for all values of r.

Chapter IX

3. $\int_0^\infty C(t)dt = \int_{2PM}^{3PM} 10^{-6}\frac{\text{gm}}{\text{liter}}\,dt = 10^{-6}\frac{\text{gm}}{\text{liter}}\text{ hour}$

$$F = \frac{Q}{\int_0^\infty C(t)dt} = \frac{1 \text{ gm}}{10^{-6}\frac{\text{gm hour}}{\text{liter}}} = 10^6\frac{\text{liter}}{\text{hour}}$$

$$V = F\frac{\int_{2PM}^{3PM} tC(t)\,dt}{\int_{2PM}^{3PM} C(t)dt} = \frac{5}{2} \times 10^6 \text{ liters}$$

4. $F = 0.0025$ cm^3/sec, $\quad V = 0.075$ cm^3

5,6. If one has enough data points to determine accurately the "tail" of the curve, the manual peel-back process can be reasonably accurate. The danger, however, lies in the possibility of a hidden component. In general, manual peel-back should be used only for quick and crude analysis.

APPENDIX V

Additional Problems

Chapter IV

1. A colony of 100 bacteria is put in a medium containing 10^{-3} grams of nutrient. Each bacterium consumes 7×10^{-12} grams of nutrient per hour. It is found that the population doubles every hour. On the assumption that this continues until all of the nutrient is gone, when is the nutrient gone? Write differential equations that describe the number of bacteria at any time. Then write and solve the differential equation that describes the amount of nutrient at any time. Give numerical answers.
2. Solve by Laplace transform

$$\frac{d^2y}{dt^2} + a\frac{dy}{dt} = 2\,e^{-at}(1-at)$$

where

$$y(0) = 0 \quad \text{and} \quad \frac{dy(0)}{dt} = 0$$

You will need the convolution formula.
3. Find the inverse transform of $g(s)$ where $g(s) = g_1(s)g_2(s)g_3(s)$ and

$$g_1(s) = \frac{1}{(s+a)^2}, \qquad g_2(s) = \frac{s}{(s+a)^2}, \qquad g_3(s) = \frac{1}{s+a}$$

Chapter V

1. Consider a three-compartment problem in which each compartment borders on the other two. Write the differential equations that describe the

concentrations as a function of time. Do not solve. What are the steady-state concentrations if the initial concentrations and volumes are $C_1(0)$, $C_2(0)$, $C_3(0)$, V_1, V_2, and V_3? Let K_A liters/sec represent the diffusion constant between compartments 1 and 2, K_B liters/sec the diffusion constant between compartments 1 and 3, and K_C liters/sec the diffusion constant between compartments 2 and 3.

2. A substance enters compartment 1 at a rate R moles/sec. It is lost from compartment 1 at a rate L_1C_1 plus diffusion into the second compartment, which occurs at a rate $K(C_1 - C_2)$. Loss from the second compartment occurs at a rate L_2C_2. Write the steady-state concentrations for the problem as given. Now suppose that at time zero K becomes zero. Find $C_1(t)$ and $C_2(t)$ after time zero.

3. Let C_{in} be the concentration of a substance inside the cell. Let C_{ext} be the concentration outside and be constant. The cell is initially in equilibrium with the outside. At time zero a pump starts operating which pumps out at a rate R moles/sec. The substance diffuses in at a rate of K moles/sec/mole/liter concentration difference.

(a) Write the differential equation and accessory conditions that describe this system after time zero. Include dimensions.

(b) What is $C_{\text{in}}(\infty)$?

(c) Guess $C_{\text{in}}(t)$. Do not verify it unless the answer to part (d) does not agree with your guess.

(d) Solve by Laplace transform. If this does not agree with part (c), verify one or the other solution. Be sure to check accessory conditions, as well as the differential equation and dimensions.

4. A system has a unit impulse response that is constant from time zero to 3 sec and becomes zero thereafter. The input is a series of six unit impulses at 1-sec intervals beginning at time zero. Describe the output.

5. Write the differential equations and accessory conditions describing the three-compartment system in which K_1 is the diffusion constant between compartments 1 and 3 and K_2 the diffusion constant between 2 and 3 (Figure AV-1). $C_1(0) = C_2(0) = 0$ and $C_3(0) = 2$ mM/liter. Do not attempt to solve them. A general solution is of course possible, but it is difficult and the result sufficiently complicated that one cannot get much feeling about the nature of the diffusion process from it. However, one can get reasonably simple solutions by making suitable approximations that will cover many interesting cases. For each of the following special cases, describe the physical nature of the approximations and write down a quick solution for $C_1(t)$, $C_2(t)$, and $C_3(t)$. What are the final concentrations in each case?

Case 1. K_1 is very big compared to K_2

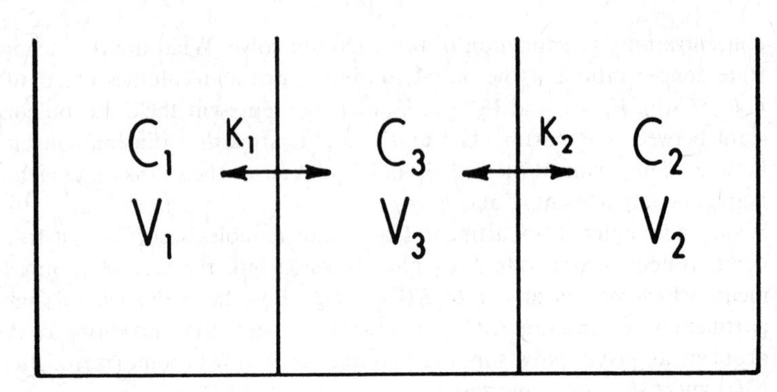

Figure AV-1

Case 2. $K_1 = K_2$ and $V_1 = V_2$

Case 3. V_2 is very small compared to V_1 and V_3, where V_2 is the extreme right-hand compartment.

Case 4. K_1 is not much different from K_2, but V_2 is very large compared to V_1 and V_3.

Chapter VII

1. The height of water in a tank can be regulated to be insensitive to changing rates of input by means of the device shown in Figure AV-2. Assume that the rate of outflow is given by the graph. Over what range of input rates would the regulation be effective? Assume that R_{in1} and R_{in2} are both within the regulating range and that R_{in2} is greater than R_{in1}. Find the change in volume V as a function of time if the input rate changes suddenly from R_{in1} to R_{in2}.

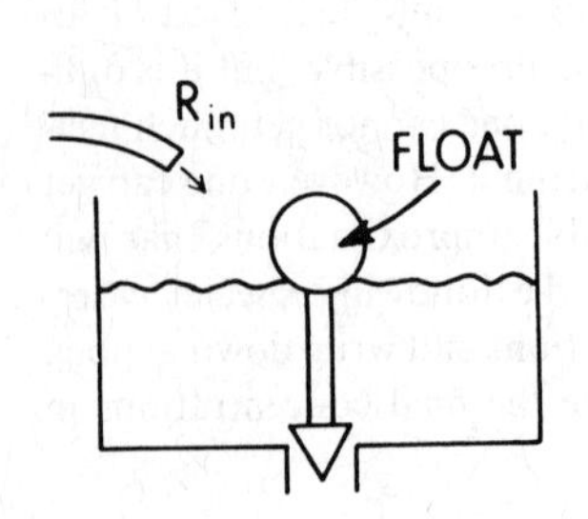

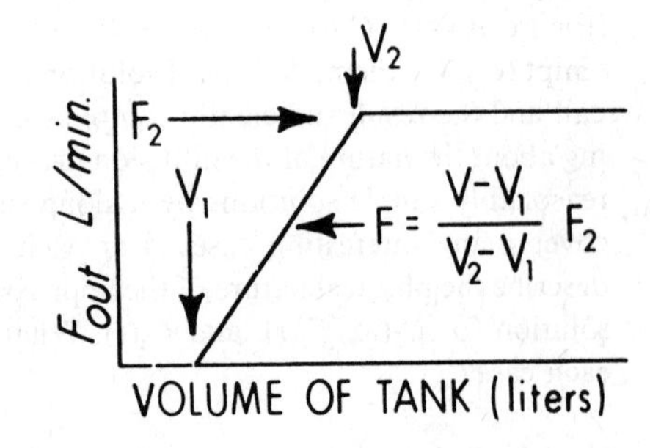

Figure AV-2

Chapter VIII

1. Consider a slab of tissue extending from $-D$ to $+D$ in which a diffusing substance is being consumed at a fixed rate R moles/cm³ sec. The tissue has a diffusion constant K. On each surface extending from $\pm D$ to $\pm (D+\delta)$ is a layer of thickness δ, in which the diffusion constant is k and the substance is not consumed. The whole thing is put in a bath of concentration C_{ex}. (See Figure AV-3.) Find the steady-state concentrations for small values of R. What is the largest value of R for which this solution is valid?
2. Two cubes of metal 10 cm on a side are connected by a glass plate of thickness 0.5 cm (Figure AV-4). At time zero the left block is at temperature 100°C and the right block at 0°C. Describe the temperature within the glass plate as a function of x and t. Assume that the initial conditions have held long enough so that the system has reached a steady state before time zero. Find an approximate solution based on the assumption that the diffusion constant in the metal is much larger than that in the glass.
   Let K_m be the diffusion constant in the metal in cal/deg cm sec
   K_g be the diffusion constant in the glass in cal/deg cm sec
   ρ_m be the density of the metal in gm/cm³
   ρ_g be the density of the glass in gm/cm³
   σ_m be the specific heat of the metal in cal/deg gm
   σ_g be the specific heat of the glass in cal/deg gm
   Assume that $\rho_m\sigma_m$ is much larger than $\rho_g\sigma_g$.

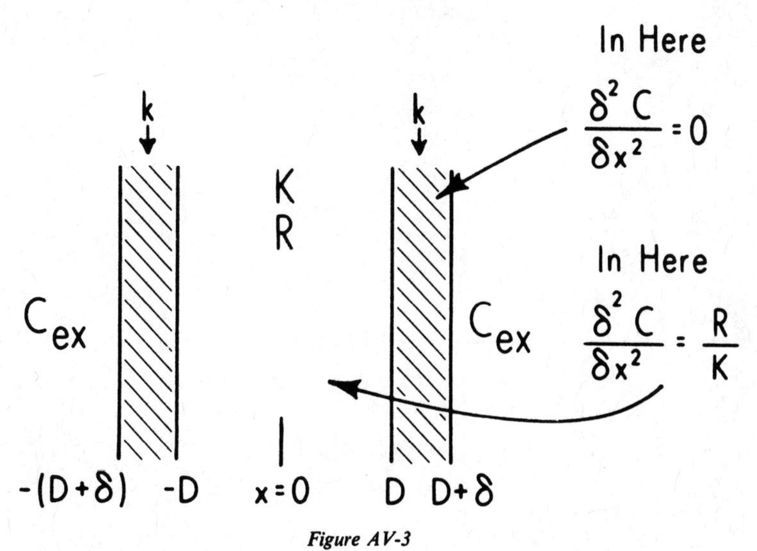

Figure AV-3

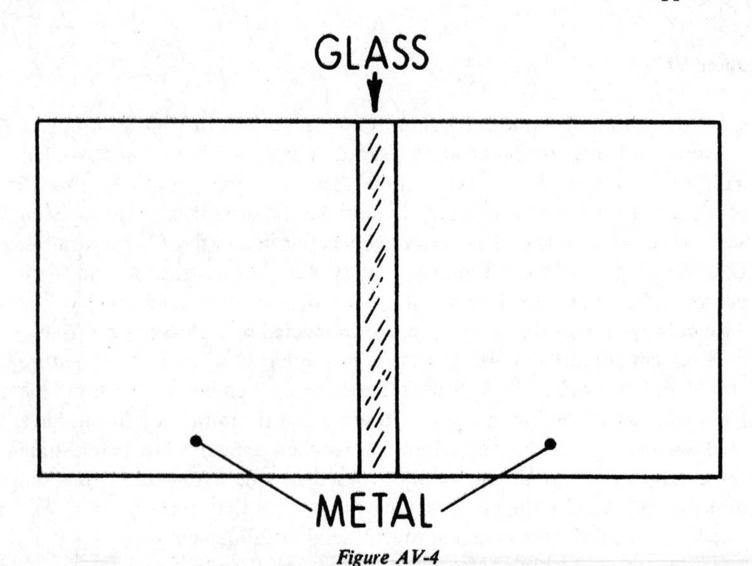

Figure AV-4

Hint: Get an approximate time-dependent solution for the temperature of the two blocks by considering them as a two-compartment diffusion problem. Then use these values as boundary conditions for the steady-state diffusion equation within the glass.

APPENDIX VI

Solutions to Additional Problems

Chapter IV

1. 20 hours

2. $y = t^2\, e^{-at}$

3. $\frac{e^{-at}}{6}\left(t^3 - \frac{at^4}{4}\right)$

Chapter V

1. $V_1 \frac{dC_1}{dt} = K_A(C_2 - C_1) + K_B(C_3 - C_1)$

$V_2 \frac{dC_2}{dt} = K_A(C_1 - C_2) + K_C(C_3 - C_2)$

$V_3 \frac{dC_3}{dt} = K_C(C_2 - C_3) + K_B(C_1 - C_3)$

In the steady state,

$$C_1 = C_2 = C_3 = \frac{V_1C_1(0) + V_2C_2(0) + V_3C_3(0)}{V_1 + V_2 + V_3}$$

2. $C_{1ss} = \frac{R(K + L_2)}{L_1L_2 + K(L_1 + L_2)}$

$C_{2ss} = \frac{RK}{L_1L_2 + K(L_1 + L_2)}$

After K becomes zero at time zero, the two compartments act independently.

$$V_1 \frac{dC_1}{dt} = R - L_1 C_1$$

$$V_2 \frac{dC_2}{dt} = -\ L_2 C_2$$

$$C_1(t) = C_{1ss} e^{-(L_1/V_1)t} + \frac{R}{L_1}(1 - e^{-(L_1/V_1)t})$$

$$C_2(t) = C_{2ss} e^{-(L_2/V_2)t}$$

3. (a) $V \frac{dC_{\text{in}}}{dt} = K(C_{\text{ext}} - C_{\text{in}}) - R \quad \text{where } C_{\text{in}}(0) = C_{\text{ext}}(0)$

(b) $C_{\text{in}}(\infty) = C_{\text{ext}} - \frac{R}{K}$

(c, d) $C_{\text{in}}(t) = C_{\text{ext}} - \frac{R}{K}(1 - e^{-(K/V)t})$

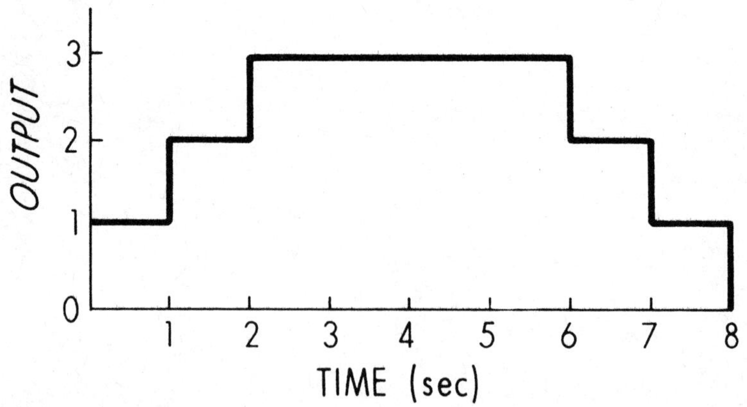

5. **Case 1.** Since K_1 is very large compared to K_2, the important effect at first is the diffusion of the contents of compartment 3 into compartment 1. For the purposes of evaluating the effects of compartments 1 and 3, compartment 2 may be neglected, but for the purposes of evaluating the contents of compartment 2, it must be regarded as filling from a time-varying compartment 3.

$$V_1 \frac{dC_1}{dt} = K(C_3 - C_1) = -\ V_3 \frac{dC_3}{dt}$$

$$C_3(t) = \frac{C_3(0)}{V_1 + V_3}[V_3 + V_1 e^{-(1/V_1+1/V_3)K_1 t}]$$

$$V_2 \frac{dC_2}{dt} = K_2 C_3(t)$$

$$C_2(t) = \frac{K_2}{V_2}\left[\frac{C_3(0)}{V_1 + V_3} V_3 t + V_1 \frac{1 - e^{-(1/V_1+1/V_3)K_1 t}}{\left(\frac{1}{V_1} + \frac{1}{V_3}\right)K_1}\right]$$

At longer times, compartments 1 and 3 can be considered in equilibrium with each other and exchanging with compartment 2. Let $V_4 = V_1 + V_3$ and

$$C_4(0) = \frac{C_3(0)V_3}{V_1 + V_3}$$

Then

$$C_2(t) = \frac{V_4 C_4(0)}{V_2 + V_4}[1 - e^{-(1/V_2+1/V_4)K_2 t}]$$

Case 2. In this case, since $V_1 = V_2$ and $K_1 = K_2$, compartments 1 and 2 can be regarded as a single compartment of volume $2V_1$, exchanging with compartment 3.
Case 3. V_2 is very small compared with V_1 and V_3. Once again, from the point of view of compartments 1 and 3, compartment 2 may be neglected. From the point of view of compartment 2, it is being filled from a time-varying compartment 3.
Case 4. V_2 is very large compared with V_1 and V_3. In this case, the build-up of concentration in compartment 2 must be very slow, and so far as compartments 1 and 3 are concerned, compartment 2 can be considered to have zero concentration.

Chapter VII

1. Regulation would be effective between input rates of zero and F_2. If the system has been in a steady state R_{in1} for some time, then the output flow must be equal to R_{in1} and the volume given by

$$R_{in1} = \frac{V - V_1}{V_2 - V_1}F_2, \qquad V(0) = V_1 + (V_2 - V_1)\frac{R_{in1}}{F_2}$$

At time zero, the input rate changes to R_{in2} so that the following differential equation describes the rate of change of volume in the tank:

$$\frac{dV}{dt} = R_{in2} - \frac{V - V_1}{V_2 - V_1} F_2$$

to which the solution is

$$V(t) = V(0)e^{-[F_2/(V_2-V_1)]t} + V(\infty)(1 - e^{-[F_2/(V_2-V_1)]t})$$

where

$$V(\infty) = V_1 + (V_2 - V_1)\frac{R_{\text{in}2}}{F_2}$$

Chapter VIII

1. Within the center slab of tissue,

$$C(X) = \frac{Rx^2}{2K} + Ax + B$$

where A and B are as yet unknown. Considering symmetry, we know that $C(D) = C(-D)$ and thus $A = 0$. Within the layer between D and $D + \delta$,

$$C(x) = \frac{x - D}{\delta}[C_{\text{ex}} - C(D)] + C(D)$$

$C(D)$ is now found from the condition that the rate of flow into the layer must be the same as that through the tissue:

$$K\frac{\partial C}{\partial x}\bigg|_{x=D\text{ in slab}} = k\frac{\partial C}{\partial x}\bigg|_{x=D\text{ in layer}}$$

$$K\frac{R}{K}D = k\frac{1}{\delta}[C_{\text{ex}} - C(D)]$$

$$C(D) = C_{\text{ex}} - \frac{RD\delta}{k}$$

The largest value of R for which this solution is valid is that for which the concentration becomes zero in the center.

2. An approximate solution of this problem depends on the fact that the diffusion of heat in metal is very rapid compared to the diffusion of heat in glass. As a result, we may neglect differences in temperature between different parts of a single block and consider the temperature of each block to change as a whole. The rate of change of temperature of a block is determined by finding the rate at which heat will cross the slab of glass given the temperatures at the two sides of the glass. Although not rigorously true, the rate of heat flow across the glass can be approximated by the solution to the steady-state problem given the temperature on the two sides, which is

$$\frac{dQ}{dt} = K_g \frac{A}{\delta}(T_1 - T_2)$$

where δ is the thickness of the glass and A the area of each face. Therefore the rate at which the hotter block cools is given by:

$$\frac{dT_1}{dt} = \frac{-1}{\rho_m \sigma_m V}\frac{dQ}{dt} = \frac{-K_g A}{\rho_m \sigma_m V\delta}(T_1 - T_2)$$

where V is the volume of one metal block.
The rate at which the temperature of the cooler block rises is given by

$$\frac{dT_2}{dt} = \frac{K_g A}{\rho_m \sigma_m V\delta}(T_1 - T_2)$$

By subtracting one equation from the other we get

$$\frac{d(T_1 - T_2)}{dt} = \frac{-2K_g A}{\rho_m \sigma_m V\delta}(T_1 - T_2)$$

$$T_1 - T_2 = [T_1(0) - T_2(0)]e^{-(2K_g A/\rho_m \sigma_m V\delta)t}.$$

Adding the same two equations we have

$$\frac{d(T_1 + T_2)}{dt} = 0$$

$$T_1 + T_2 = T_1(0) + T_2(0)$$

$$T_1(t) = \frac{1}{2}T_1(0)(1 + e^{-(2K_g A/\rho_m \sigma_m V\delta)t})$$

$$T_2(t) = \frac{1}{2}T_1(0)(1 - e^{-(2K_g A/\rho_m \sigma_m V\delta)t})$$

Using these values we may now determine the temperature within the glass as a function of space and time:

$$T(x, t) = \left(1 - \frac{x}{\delta}\right)T_1(t) + \frac{x}{\delta}T_2(t)$$

APPENDIX VII

Sample Programs

This appendix contains a number of sample programs illustrating the techniques described in Chapter VI, "Numerical Methods." The sample programs are given in Microsoft BASIC, but will run in most versions of BASIC. Graphics instructions, as of this writing, are not standardized. The ones illustrated are for IBM PCs, but modification for other computers is quite straightforward.*

Throughout this book, emphasis is placed upon keeping track of the dimensions of physical quantities as they are incorporated into equations and as they appear in the answers to problems. Unfortunately, most computer languages do not make use of dimensions. In principle, one could keep track of dimensions by means of reminder (REM) statements, but this is unduly tiresome. The practical rule of thumb is to follow the techniques as illustrated in Chapter I, in which dimensions are written (actually or mentally) beneath equations. The equations are transferred into the computer program without attempting to carry the dimensions. It should be emphasized at this point that although no dimensions are carried in the computer program, the parameters there used are not necessarily dimensionless parameters, as defined in Chapter II, Section 7.

The following program carries out the computation described in Chapter VI, Section 1.

```
30 V1=10
40 V2=20
50 FLOW=1
60 TIME=0
```

*Microsoft and BASIC are trademarks of Microsoft Corporation, and IBM and PC are trademarks of International Business Machines Corporation.

```
 70 INPUT "TYPE INITIAL C1",CINITIAL
 80 C1=CINITIAL
 90 C2=0
100 TINC=1
120 PRINT "TIME= ";TIME,"C1= ";C1,"C2= ";C2
160 C1=CINITIAL*EXP(-FLOW*TIME/V1)
170 C2INC=TINC*FLOW*(C1-C2)/V2
180 C2=C2+C2INC
190 TIME=TIME+TINC
220 GOTO 120
```

In the form given above, the program prints the time and the concentrations in the two compartments for every incremental time step. This causes the program to run fairly slowly, but more importantly, it clutters the screen with more data than one can easily read at one time. Therefore, let us modify the program so that the values are printed only every fifth step, by inserting line 125 and line 200 shown below.

```
 30 V1=10
 40 V2=20
 50 FLOW=1
 60 TIME=0
 70 INPUT "TYPE INITIAL C1",CINITIAL
 80 C1=CINITIAL
 90 C2=0
100 TINC=1
120 PRINT "TIME= ";TIME,"C1= ";C1,"C2= ";C2
125 FOR I=1 TO 5
160 C1=CINITIAL*EXP(-FLOW*TIME/V1)
170 C2INC=TINC*FLOW*(C1-C2)/V2
180 C2=C2+C2INC
190 TIME=TIME+TINC
200 NEXT I
220 GOTO 120
```

These programs are conveniently run with initial concentration of $C_1 = 100$ of whatever units of concentration you wish. This corresponds to the example shown in Figure 2-6, Chapter II.

Here is another version of the same program that produces a graphical output on an IBM PC color display.

```
 10 BASE=150
 20 KEY OFF: SCREEN 0:SCREEN 1
 30 V1=10
 40 V2=20
 50 FLOW=1
 60 TIME=0
 70 INPUT "TYPE INITIAL C1",CINITIAL
 80 C1=CINITIAL
 90 C2=0
100 TINC=1
120 X=5*TIME+30
130 PSET (X,BASE-C1):PSET (X,BASE-C2):PSET(X,BASE)
160 C1=CINITIAL*EXP(-FLOW*TIME/V1)
170 C2INC=TINC*FLOW*(C1-C2)/V2
180 C2=C2+C2INC
190 TIME=TIME+TINC
220 GOTO 120
```

The next program is an example of the reverse computation technique, described in Chapter VI, Section 2. It is identical with the previous program except that when the time limit of 90 sec is reached, the time increment is replaced by the negative of its previous value, thereby causing the program to step backwards in time. Try this program with different time increments to reproduce the results shown in Chapter VI.

```
 10 BASE=150
 20 KEY OFF:SCREEN 0:SCREEN 1: COLOR 0,1
 30 V1=10
 40 V2=20
 50 FLOW=1
 60 TIME=0
 70 CINITIAL=400:REM  THIS DIFFERS FROM TEXT EXAMPLE
 80 C1=CINITIAL
 90 C2=0
100 TINC=1
110 HUE=3
120 X=30+3*TIME
130 IF TINC<0 THEN HUE=2
140 PSET (X,BASE)
150 PSET (X,BASE-C2), HUE
160 C1=CINITIAL*EXP(-FLOW*TIME/V1)
170 C2INC=TINC*FLOW*(C1-C2)/V2
180 C2=C2+C2INC
```

```
190 TIME=TIME+TINC
200 IF TIME>90 THEN TINC=-TINC
210 IF TIME<0 THEN STOP
220 GOTO 120
```

The following program solves numerically a problem on page 123 except that R is assumed a function of position. The following parameters are used:

$$K=0.5 \qquad C_{EXT}=150$$
$$R=0.02 \qquad \text{on right side}$$
$$=0 \qquad \text{on left side}$$
$$\Delta x=1$$

The program is set up so that the full diffusion width is represented by 30 values of the x coordinate. Of these, R is assumed zero for the left 15 and equal to 0.02 for the right 15. The program runs with two different sets of first approximations as described in the first paragraph on page 124. The first set sets all of the concentration values except the endpoints to zero and successive passes through the iterative process will cause the concentrations to increase toward the final values. One hundred twenty such iterative passes are made. A new set of starting values is then entered with all of the concentrations equal to C external and the iterative process repeated another 120 times. In this case, since all of the values, except the ends, start too high, they will move down toward the last set of values associated with the zero initial trial values as described in the second paragraph on page 124.

```
 60 KEY OFF
 70 BASE=180
 80 SCREEN 0
 90 SCREEN 2
100 N=30
105 M=15
110 K=.5
120 R=.02
130 CEXT=150
150 DIM CA(N),CB(N),R(N)
160 CSTART=0
170 STEPX=400/N
210 FOR I=1 TO N:R(I)=R:NEXT I
220 FOR I=1 TO M:R(I)=0:NEXT I
230 CA(1)=CEXT
240 CA(N)=CEXT
250 FOR I=2 TO N-1:CA(I)=CSTART:NEXT I
```

```
260 FOR ITERATIONS = 1 TO 180
290 X=100
300 FOR I=1 TO N
310 X=X+STEPX
320 Y=BASE-CA(I)
330 PSET (X,BASE)
340 PSET (X,Y)
350 NEXT I
400 FOR I=2 TO N-1
410 CB(I)=.5*((-R(I)*CA(I)/K)+CA(I+1)+CA(I-1))
420 NEXT I
430 FOR I=2 TO N-1
440 CA(I)=CB(I):NEXT I
450 NEXT ITERATIONS
500 IF CSTART=CEXT THEN STOP
510 CSTART=CEXT
520 GOTO 250
```

The next program illustrates the solution of a set of simultaneous differential equations. The reader may want to deduce from the program what the differential equations and accessory conditions are. Hint: they are two first-order equations. Try combining into a single second-order differential equation and see if you recognize it. Second hint: see the introduction to Chapter VII. Third hint: run the program.

```
 80 KEY OFF
 90 SCREEN 0
100 SCREEN 1
110 C=25
120 S=0
130 X=20
140 MID=50
150 DEL=.1
190 FOR I=1 TO 200
200 PSET(X,MID),1
210 PSET(X,MID-S),2
220 PSET (X,MID-C),3
250 C=C-DEL*S
260 S=S+DEL*C
270 X=X+1
300 NEXT I
```

References

1. Bayliss, L. E., "Living Control Systems," Freeman, San Francisco, California, 1966.

2. Berman, M., Weiss, M. F., and Shahn, E., Some formal approaches to the Analysis of kinetic data in terms of linear compartmental systems, *Biophys. J.* **2**, 289–316 (1962).

3. Cardon, S. Z., and Iberall, A. S., Oscillations in biological systems, *Currents Mod. Biol.* **3**, 237–249, 1970.

4. Chance, B., and Yoshioka, T., Sustained oscillations of ionic constituents of mitochondria, *Arch. Biochem. Biophys.* **117**, 451 (1966).

5. Churchill, R. V., "Fourier Series and Boundary Value Problems," 2nd Ed., McGraw-Hill, New York, 1963.

6. Churchill, R. V., "Operational Mathematics," 2nd Ed., McGraw-Hill, New York, 1958.

7. Crank, J., "The Mathematics of Diffusion," Oxford Univ. Press, London and New York, 1964.

8. "C.R.C. Standard Mathematical Tables," Chemical Rubber Publ. Co., Cleveland, Ohio.

9. Doetsch, G., "Guide to the Applications of Laplace Transforms," Van Nostrand, New Jersey, 1961.

10. Doyle, A. C., The sign of the four. *In* "The Annotated Sherlock Holmes," Vol. 1, p. 638. Baring-Gould, Potter, New York, 1967.

11. Forster, R. E., Factors affecting the rate of exchange of O_2 between blood and tissues. *In* "Oxygen in the Animal Organism" (F. Dickens, and E. Neil, eds.), Macmillan, New York, 1964.

12. Gatewood, L. G., Ackerman, E., Rosevear, J. W., and Molnar, G. D., Modeling blood glucose dynamics, *Behavioral Sci.* **15**, 72–87, 1970.

13. Grodins, F. S., "Control Theory and Biological Systems," Columbia Univ. Press, New York, 1963.

13a. Hess, H., Boiteux, A., and Krüger, J., Cooperation of glycolytic enzymes. *In* "Advances in Enzyme Regulation" (G. Weber, ed.), Pergamon, Oxford, 1969.

14. Higgins, J., A chemical mechanism for oscillation of glycolytic intermediates in yeast cells, *Proc. Nat. Acad. Sci.* **51**, 989–994, (1964).

15. Hildebrand, F. B., "Introduction to Numerical Analysis," McGraw-Hill, New York, 1956.

16. Kalmus, H., "Regulation and Control in Living Systems," Wiley, New York, 1966.

17. Kety, S. S., and Schmidt, C. F., The nitrous oxide method for the quantitative determination of cerebral blood flow in man; theory, procedure and normal values. *J. Clin. Invest.* **27**, 476–483 (1948).

18. Kleppner, D. and Ramsey, N., "Quick Calculus," Wiley, New York, 1965.

19. Milsum, J. H., "Biological Control System Analysis," McGraw-Hill, New York, 1966.

20. Navas, F. and Stark, L., Sampling or intermittency in hand control system dynamics, *Biophys. J.* **8**, 252–302 (1968).

21. Pavlidis, T. and Kauzmann, W., Toward a quantitative biochemical model for Circadian oscillators. *Arch. Biochem. Biophys.* **132**, 338–348 (1969).

22. Reiner, J. M., "The Organism as an Adaptive Control System," Prentice-Hall, Englewood Cliffs, New Jersey, 1968.

23. Richmond, D. E., "Introductory Calculus," Addison-Wesley, Reading, Massachusetts, 1959.

24. Simon, W., A method of exponential separation applicable to small computers, *Phys. Med. Biol.* **15**, 355–360 (1970).

25. Snyder, W. S., Fisk, B. R., Bernard, S. R., Ford, M. R., and Muir, J. R., Urinary excretion of tritium following exposure of man to HTO–a two exponential model, *Phys. Med. Biol.* **13**, 547–559 (1968).

26. Sollberger, A., "Biological Rhythm Research," Elsevier, Amsterdam, 1965.

27. Solomon, A. K., Compartmental method of kinetic analysis. *In* "Mineral Metabolism" (C. L. Comar, and F. Bronner, eds.), Vol. 1, pp. 119–167, Academic Press, New York, 1960.

28. Stark, L., Vision: Servoanalysis of pupil reflex to light. *In* "Medical Physics" (O. Glasser, ed.), pp. 702–719, Year Book Publ., Chicago, Illinois, 1960.

29. Stark, L., Neurological feedback control systems. *In* "Advances in Bioengineering and Instrumentation" (F. Alt, ed.), Chap. 3, p. 289, Plenum, New York, 1966.

30. Thompson, H. E., Horgan, J. D., and Delfs, E., A simplified mathematical model and simulations of the hypophysics–ovarian endocrine control system. *Biophys. J.* **9**, 278–291 (1969).

31. Walter, C. F., "Biological Control Mechanisms," Academic Press, New York, 1969.

32. Wright, B. E., The use of kinetic models to analyze differentiation, *Behavioral Sci.* **15**, 37–45 (1970).

33. Zierler, K. L., Circulation times and the theory of indicator dilution methods for determining blood flow and volume. *In* "Handbook of Physiology," (W. F. Hamilton, and F. Dow, eds.), Vol. 1, Sec. 2, Chapter 18, American Physiological Society, Washington, D.C., 1962.

Index

A

B

C

P

Q

R

S

T

U

V

W

X

A CATALOG OF SELECTED

DOVER BOOKS

IN ALL FIELDS OF INTEREST

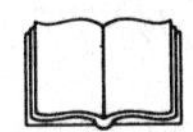

CATALOG OF DOVER BOOKS

YUCATAN BEFORE AND AFTER THE CONQUEST, Diego de Landa. Only significant account of Yucatan written in the early post-Conquest era. Translated by William Gates. Over 120 illustrations. 162pp. 5⅜ × 8½. 23622-6 Pa. **$3.50**

ORNATE PICTORIAL CALLIGRAPHY, E.A. Lupfer. Complete instructions, over 150 examples help you create magnificent "flourishes" from which beautiful animals and objects gracefully emerge. 8⅛ × 11. 21957-7 Pa. **$2.95**

DOLLY DINGLE PAPER DOLLS, Grace Drayton. Cute chubby children by same artist who did Campbell Kids. Rare plates from 1910s. 30 paper dolls and over 100 outfits reproduced in full color. 32pp. 9¼ × 12¼. 23711-7 Pa. **$3.50**

CURIOUS GEORGE PAPER DOLLS IN FULL COLOR, H. A. Rey, Kathy Allert. Naughty little monkey-hero of children's books in two doll figures, plus 48 full-color costumes: pirate, Indian chief, fireman, more. 32pp. 9¼ × 12¼. 24386-9 Pa. **$3.50**

GERMAN: HOW TO SPEAK AND WRITE IT, Joseph Rosenberg. Like *French, How to Speak and Write It.* Very rich modern course, with a wealth of pictorial material. 330 illustrations. 384pp. 5⅜ × 8½. (USUKO) 20271-2 Pa. **$4.75**

CATS AND KITTENS: 24 Ready-to-Mail Color Photo Postcards, D. Holby. Handsome collection; feline in a variety of adorable poses. Identifications. 12pp. on postcard stock. 8¼ × 11. 24469-5 Pa. **$2.95**

MARILYN MONROE PAPER DOLLS, Tom Tierney. 31 full-color designs on heavy stock, from *The Asphalt Jungle, Gentlemen Prefer Blondes,* 22 others. 1 doll. 16 plates. 32pp. 9⅜ × 12¼. 23769-9 Pa. **$3.50**

FUNDAMENTALS OF LAYOUT, F.H. Wills. All phases of layout design discussed and illustrated in 121 illustrations. Indispensable as student's text or handbook for professional. 124pp. 8⅛ × 11. 21279-3 Pa. **$4.50**

FANTASTIC SUPER STICKERS, Ed Sibbett, Jr. 75 colorful pressure-sensitive stickers. Peel off and place for a touch of pizzazz: clowns, penguins, teddy bears, etc. Full color. 16pp. 8¼ × 11. 24471-7 Pa. **$2.95**

LABELS FOR ALL OCCASIONS, Ed Sibbett, Jr. 6 labels each of 16 different designs—baroque, art nouveau, art deco, Pennsylvania Dutch, etc.—in full color. 24pp. 8¼ × 11. 23688-9 Pa. **$2.95**

HOW TO CALCULATE QUICKLY: RAPID METHODS IN BASIC MATHEMATICS, Henry Sticker. Addition, subtraction, multiplication, division, checks, etc. More than 8000 problems, solutions. 185pp. 5 × 7¼. 20295-X Pa. **$2.95**

THE CAT COLORING BOOK, Karen Baldauski. Handsome, realistic renderings of 40 splendid felines, from American shorthair to exotic types. 44 plates. Captions. 48pp. 8¼ × 11. 24011-8 Pa. **$2.25**

THE TALE OF PETER RABBIT, Beatrix Potter. The inimitable Peter's terrifying adventure in Mr. McGregor's garden, with all 27 wonderful, full-color Potter illustrations. 55pp. 4¼ × 5½. (Available in U.S. only) 22827-4 Pa. **$1.75**

BASIC ELECTRICITY, U.S. Bureau of Naval Personnel. Batteries, circuits, conductors, AC and DC, inductance and capacitance, generators, motors, transformers, amplifiers, etc. 349 illustrations. 448pp. 6½ × 9¼. 20973-3 Pa. **$7.95**

REASON IN ART, George Santayana. Renowned philosopher's provocative, seminal treatment of basis of art in instinct and experience. Volume Four of *The Life of Reason*. 230pp. 5⅜ × 8. 24358-3 Pa. $4.50

LANGUAGE, TRUTH AND LOGIC, Alfred J. Ayer. Famous, clear introduction to Vienna, Cambridge schools of Logical Positivism. Role of philosophy, elimination of metaphysics, nature of analysis, etc. 160pp. 5⅜ × 8½. (USCO) 20010-8 Pa. $2.75

BASIC ELECTRONICS, U.S. Bureau of Naval Personnel. Electron tubes, circuits, antennas, AM, FM, and CW transmission and receiving, etc. 560 illustrations. 567pp. 6½ × 9¼. 21076-6 Pa. $8.95

THE ART DECO STYLE, edited by Theodore Menten. Furniture, jewelry, metalwork, ceramics, fabrics, lighting fixtures, interior decors, exteriors, graphics from pure French sources. Over 400 photographs. 183pp. 8⅜ × 11¼. 22824-X Pa. $6.95

THE FOUR BOOKS OF ARCHITECTURE, Andrea Palladio. 16th-century classic covers classical architectural remains, Renaissance revivals, classical orders, etc. 1738 Ware English edition. 216 plates. 110pp. of text. 9½ × 12¾. 21308-0 Pa. $11.50

THE WIT AND HUMOR OF OSCAR WILDE, edited by Alvin Redman. More than 1000 ripostes, paradoxes, wisecracks: Work is the curse of the drinking classes, I can resist everything except temptations, etc. 258pp. 5⅜ × 8½. (USCO) 20602-5 Pa. $3.95

THE DEVIL'S DICTIONARY, Ambrose Bierce. Barbed, bitter, brilliant witticisms in the form of a dictionary. Best, most ferocious satire America has produced. 145pp. 5⅜ × 8½. 20487-1 Pa. $2.50

ERTÉ'S FASHION DESIGNS, Erté. 210 black-and-white inventions from *Harper's Bazar*, 1918-32, plus 8pp. full-color covers. Captions. 88pp. 9 × 12. 24203-X Pa. $6.50

ERTÉ GRAPHICS, Erté. Collection of striking color graphics: *Seasons, Alphabet, Numerals, Aces* and *Precious Stones*. 50 plates, including 4 on covers. 48pp. 9⅜ × 12¼. 23580-7 Pa. $6.95

PAPER FOLDING FOR BEGINNERS, William D. Murray and Francis J. Rigney. Clearest book for making origami sail boats, roosters, frogs that move legs, etc. 40 projects. More than 275 illustrations. 94pp. 5⅜ × 8½. 20713-7 Pa. $2.25

ORIGAMI FOR THE ENTHUSIAST, John Montroll. Fish, ostrich, peacock, squirrel, rhinoceros, Pegasus, 19 other intricate subjects. Instructions. Diagrams. 128pp. 9 × 12. 23799-0 Pa. $4.95

CROCHETING NOVELTY POT HOLDERS, edited by Linda Macho. 64 useful, whimsical pot holders feature kitchen themes, animals, flowers, other novelties. Surprisingly easy to crochet. Complete instructions. 48pp. 8¼ × 11. 24296-X Pa. $1.95

CROCHETING DOILIES, edited by Rita Weiss. Irish Crochet, Jewel, Star Wheel, Vanity Fair and more. Also luncheon and console sets, runners and centerpieces. 51 illustrations. 48pp. 8¼ × 11. 23424-X Pa. $2.50

THE RIME OF THE ANCIENT MARINER, Gustave Doré, S.T. Coleridge. Doré's finest work, 34 plates capture moods, subtleties of poem. Full text. 77pp. 9¼ × 12. 22305-1 Pa. $4.95

SONGS OF INNOCENCE, William Blake. The first and most popular of Blake's famous "Illuminated Books," in a facsimile edition reproducing all 31 brightly colored plates. Additional printed text of each poem. 64pp. 5¼ × 7. 22764-2 Pa. $3.50

AN INTRODUCTION TO INFORMATION THEORY, J.R. Pierce. Second (1980) edition of most impressive non-technical account available. Encoding, entropy, noisy channel, related areas, etc. 320pp. 5⅜ × 8½. 24061-4 Pa. $4.95

THE DIVINE PROPORTION: A STUDY IN MATHEMATICAL BEAUTY, H.E. Huntley. "Divine proportion" or "golden ratio" in poetry, Pascal's triangle, philosophy, psychology, music, mathematical figures, etc. Excellent bridge between science and art. 58 figures. 185pp. 5⅜ × 8½. 22254-3 Pa. $3.95

THE DOVER NEW YORK WALKING GUIDE: From the Battery to Wall Street, Mary J. Shapiro. Superb inexpensive guide to historic buildings and locales in lower Manhattan: Trinity Church, Bowling Green, more. Complete Text; maps. 36 illustrations. 48pp. 3⅞ × 9¼. 24225-0 Pa. $2.50

NEW YORK THEN AND NOW, Edward B. Watson, Edmund V. Gillon, Jr. 83 important Manhattan sites: on facing pages early photographs (1875-1925) and 1976 photos by Gillon. 172 illustrations. 171pp. 9¼ × 10. 23361-8 Pa. $7.95

HISTORIC COSTUME IN PICTURES, Braun & Schneider. Over 1450 costumed figures from dawn of civilization to end of 19th century. English captions. 125 plates. 256pp. 8⅜ × 11¼. 23150-X Pa. $7.50

VICTORIAN AND EDWARDIAN FASHION: A Photographic Survey, Alison Gernsheim. First fashion history completely illustrated by contemporary photographs. Full text plus 235 photos, 1840-1914, in which many celebrities appear. 240pp. 6½ × 9¼. 24205-6 Pa. $6.00

CHARTED CHRISTMAS DESIGNS FOR COUNTED CROSS-STITCH AND OTHER NEEDLECRAFTS, Lindberg Press. Charted designs for 45 beautiful needlecraft projects with many yuletide and wintertime motifs. 48pp. 8¼ × 11. 24356-7 Pa. $2.50

101 FOLK DESIGNS FOR COUNTED CROSS-STITCH AND OTHER NEEDLECRAFTS, Carter Houck. 101 authentic charted folk designs in a wide array of lovely representations with many suggestions for effective use. 48pp. 8¼ × 11. 24369-9 Pa. $2.25

FIVE ACRES AND INDEPENDENCE, Maurice G. Kains. Great back-to-the-land classic explains basics of self-sufficient farming. The one book to get. 95 illustrations. 397pp. 5⅜ × 8½. 20974-1 Pa. $4.95

A MODERN HERBAL, Margaret Grieve. Much the fullest, most exact, most useful compilation of herbal material. Gigantic alphabetical encyclopedia, from aconite to zedoary, gives botanical information, medical properties, folklore, economic uses, and much else. Indispensable to serious reader. 161 illustrations. 888pp. 6½ × 9¼. (Available in U.S. only) 22798-7, 22799-5 Pa., Two-vol. set $16.45

DECORATIVE NAPKIN FOLDING FOR BEGINNERS, Lillian Oppenheimer and Natalie Epstein. 22 different napkin folds in the shape of a heart, clown's hat, love knot, etc. 63 drawings. 48pp. 8¼ × 11. 23797-4 Pa. $1.95

DECORATIVE LABELS FOR HOME CANNING, PRESERVING, AND OTHER HOUSEHOLD AND GIFT USES, Theodore Menten. 128 gummed, perforated labels, beautifully printed in 2 colors. 12 versions. Adhere to metal, glass, wood, ceramics. 24pp. 8¼ × 11. 23219-0 Pa. $2.95

EARLY AMERICAN STENCILS ON WALLS AND FURNITURE, Janet Waring. Thorough coverage of 19th-century folk art: techniques, artifacts, surviving specimens. 166 illustrations, 7 in color. 147pp. of text. 7⅞ × 10¾. 21906-2 Pa. $9.95

AMERICAN ANTIQUE WEATHERVANES, A.B. & W.T. Westervelt. Extensively illustrated 1883 catalog exhibiting over 550 copper weathervanes and finials. Excellent primary source by one of the principal manufacturers. 104pp. 6⅛ × 9¼. 24396-6 Pa. $3.95

ART STUDENTS' ANATOMY, Edmond J. Farris. Long favorite in art schools. Basic elements, common positions, actions. Full text, 158 illustrations. 159pp. 5⅜ × 8½. 20744-7 Pa. $3.95

BRIDGMAN'S LIFE DRAWING, George B. Bridgman. More than 500 drawings and text teach you to abstract the body into its major masses. Also specific areas of anatomy. 192pp. 6½ × 9¼. (EA) 22710-3 Pa. $4.50

COMPLETE PRELUDES AND ETUDES FOR SOLO PIANO, Frederic Chopin. All 26 Preludes, all 27 Etudes by greatest composer of piano music. Authoritative Paderewski edition. 224pp. 9 × 12. (Available in U.S. only) 24052-5 Pa. $7.50

PIANO MUSIC 1888-1905, Claude Debussy. Deux Arabesques, Suite Bergamesque, Masques, 1st series of Images, etc. 9 others, in corrected editions. 175pp. 9⅜ × 12¼. (ECE) 22771-5 Pa. $5.95

TEDDY BEAR IRON-ON TRANSFER PATTERNS, Ted Menten. 80 iron-on transfer patterns of male and female Teddys in a wide variety of activities, poses, sizes. 48pp. 8¼ × 11. 24596-9 Pa. $2.25

A PICTURE HISTORY OF THE BROOKLYN BRIDGE, M.J. Shapiro. Profusely illustrated account of greatest engineering achievement of 19th century. 167 rare photos & engravings recall construction, human drama. Extensive, detailed text. 122pp. 8¼ × 11. 24403-2 Pa. $7.95

NEW YORK IN THE THIRTIES, Berenice Abbott. Noted photographer's fascinating study shows new buildings that have become famous and old sights that have disappeared forever. 97 photographs. 97pp. 11⅜ × 10. 22967-X Pa. $7.50

MATHEMATICAL TABLES AND FORMULAS, Robert D. Carmichael and Edwin R. Smith. Logarithms, sines, tangents, trig functions, powers, roots, reciprocals, exponential and hyperbolic functions, formulas and theorems. 269pp. 5⅜ × 8½. 60111-0 Pa. $4.95

HANDBOOK OF MATHEMATICAL FUNCTIONS WITH FORMULAS, GRAPHS, AND MATHEMATICAL TABLES, edited by Milton Abramowitz and Irene A. Stegun. Vast compendium: 29 sets of tables, some to as high as 20 places. 1,046pp. 8 × 10½. 61272-4 Pa. $19.95

TWENTY-FOUR ART NOUVEAU POSTCARDS IN FULL COLOR FROM CLASSIC POSTERS, Hayward and Blanche Cirker. Ready-to-mail postcards reproduced from rare set of poster art. Works by Toulouse-Lautrec, Parrish, Steinlen, Mucha, Cheret, others. 12pp. 8¼× 11. 24389-3 Pa. $2.95

READY-TO-USE ART NOUVEAU BOOKMARKS IN FULL COLOR, Carol Belanger Grafton. 30 elegant bookmarks featuring graceful, flowing lines, foliate motifs, sensuous women characteristic of Art Nouveau. Perforated for easy detaching. 16pp. 8¼ × 11. 24305-2 Pa. $2.95

FRUIT KEY AND TWIG KEY TO TREES AND SHRUBS, William M. Harlow. Fruit key covers 120 deciduous and evergreen species; twig key covers 160 deciduous species. Easily used. Over 300 photographs. 126pp. 5⅜ × 8½. 20511-8 Pa. $2.25

LEONARDO DRAWINGS, Leonardo da Vinci. Plants, landscapes, human face and figure, etc., plus studies for Sforza monument, *Last Supper*, more. 60 illustrations. 64pp. 8¼ × 11⅝. 23951-9 Pa. $2.75

CLASSIC BASEBALL CARDS, edited by Bert R. Sugar. 98 classic cards on heavy stock, full color, perforated for detaching. Ruth, Cobb, Durocher, DiMaggio, H. Wagner, 99 others. Rare originals cost hundreds. 16pp. 8¼ × 11. 23498-3 Pa. $3.25

TREES OF THE EASTERN AND CENTRAL UNITED STATES AND CANADA, William M. Harlow. Best one-volume guide to 140 trees. Full descriptions, woodlore, range, etc. Over 600 illustrations. Handy size. 288pp. 4½ × 6⅜. 20395-6 Pa. $3.95

JUDY GARLAND PAPER DOLLS IN FULL COLOR, Tom Tierney. 3 Judy Garland paper dolls (teenager, grown-up, and mature woman) and 30 gorgeous costumes highlighting memorable career. Captions. 32pp. 9¼ × 12¼. 24404-0 Pa. $3.50

GREAT FASHION DESIGNS OF THE BELLE EPOQUE PAPER DOLLS IN FULL COLOR, Tom Tierney. Two dolls and 30 costumes meticulously rendered. Haute couture by Worth, Lanvin, Paquin, other greats late Victorian to WWI. 32pp. 9¼ × 12¼. 24425-3 Pa. $3.50

FASHION PAPER DOLLS FROM GODEY'S LADY'S BOOK, 1840-1854, Susan Johnston. In full color: 7 female fashion dolls with 50 costumes. Little girl's, bridal, riding, bathing, wedding, evening, everyday, etc. 32pp. 9¼ × 12¼. 23511-4 Pa. $3.95

THE BOOK OF THE SACRED MAGIC OF ABRAMELIN THE MAGE, translated by S. MacGregor Mathers. Medieval manuscript of ceremonial magic. Basic document in Aleister Crowley, Golden Dawn groups. 268pp. 5⅜ × 8½. 23211-5 Pa. $5.00

PETER RABBIT POSTCARDS IN FULL COLOR: 24 Ready-to-Mail Cards, Susan Whited LaBelle. Bunnies ice-skating, coloring Easter eggs, making valentines, many other charming scenes. 24 perforated full-color postcards, each measuring 4¼ × 6, on coated stock. 12pp. 9 × 12. 24617-5 Pa. $2.95

CELTIC HAND STROKE BY STROKE, A. Baker. Complete guide creating each letter of the alphabet in distinctive Celtic manner. Covers hand position, strokes, pens, inks, paper, more. Illustrated. 48pp. 8¼ × 11. 24336-2 Pa. $2.50